Preparing the Ground for Renewal of Nuclear Power

Preparing the Ground for Renewal of Nuclear Power

Edited by

Behram N. Kursunoglu
Global Foundation, Inc.
Coral Gables, Florida

Stephan L. Mintz
Florida International University
Miami, Florida

and

Arnold Perlmutter
University of Miami
Coral Gables, Florida

Springer Science+Business Media, LLC

Proceedings of the International Conference on Preparing the Ground for Renewal of Nuclear Power, held October 22–23, 1998, in Paris, France

ISBN 978-1-4613-7118-2 ISBN 978-1-4615-4679-5 (eBook)
DOI 10.1007/978-1-4615-4679-5

Originally published by Kluwer Academic / Plenum Publishers, New York in 1999
Softcover reprint of the hardcover 1st edition 1999

10 9 8 7 6 5 4 3 2 1

A C.I.P. record for this book is available from the Library of Congress.

Preface

The *International Conference on Preparing the Ground for Renewal of Nuclear Power* convened from October 22-23, 1998 in Paris, France. Like its precedents (1979 in Nice, France; 1982 and 1989 in Paris, France), the conference was very successful. It was organized by the European Subcommittee of the Global Foundation under the direction of Jean Couture (Trustee of the Foundation) and his most able colleague Pierre Zaleski (member of the Global Foundation Advisory Board). They chose the topic and consulted with the Global Foundation in the United States. The topic is of greatest importance for the global energy problem that carries with it certain problematic issues like waste and possible proliferation of nuclear weapons. These problems have been addressed in the past in this same series of conferences which began in 1977 under the title *International Scientific Forum on Energy.* The above problems do still constitute a challenge. Despite these shortcomings, nuclear energy is an indispensable source necessary for meeting the global demand as well as avoiding global warming generated by the use of fossil fuels, the consequences of which are irreversible.

The French obtain more than eighty percent of their electrical energy from nuclear sources which can be compared to the United States' reliance on only eighteen percent of total energy consumption. The spirits of the approach we have taken in this series of conferences will continue in our 1999 Washington, D.C. meeting from November 5-6. We hope to make further progress.

We would like to take this opportunity to record our appreciation to the Electric Power Research Institute (EPRI) for their continued financial and intellectual support of this series from its inception.

Behram N. Kursunoglu
Stephan L. Mintz
Arnold Perlmutter

About the Global Foundation, Inc.

The Global Foundation, Inc. was established in 1977 and utilizes the world's most important resource . . . people. The Foundation consists of senior men and women in science and learning; outstanding achievers and entrepreneurs from industry, governments, and international organizations; and promising, enthusiastic young people. These people form a unique and distinguished interdisciplinary entity, and the Foundation is dedicated to assembling all the resources necessary for them to work together. The distinguished senior members of the Foundation convey their expertise and accumulated experience, knowledge, and wisdom to the younger membership on important global issues and frontier problems in science.

Our work is a common effort, employing the ideas of creative thinkers with a wide range of experiences and viewpoints.

Global Foundation Board of Trustees

Global Foundation's Recent Conference Proceedings

Making the Market Right for the Efficient Use of Energy
Edited by: Behram N. Kursunoglu
Nova Science Publishers, Inc., New York, 1992

Unified Symmetry in the Small and in the Large
Edited by: Behram N. Kursunoglu and Arnold Perlmutter
Nova Science Publishers, Inc., New York, 1993

Unified Symmetry in the Small and in the Large - 1
Edited by: Behram N. Kursunoglu, Stephan Mintz and Arnold Perlmutter
Plenum Press, 1994.

Unified Symmetry in the Small and in the Large - 2
Edited by: Behram N. Kursunoglu, Stephan Mintz and Arnold Perlmutter
Plenum Press, 1995.

Global Energy Demand in Transition: The New Role of Electricity
Edited by: Behram N. Kursunoglu, Stephan Mintz and Arnold Perlmutter
Plenum Press, 1996.

Economics and Politics of Energy
Edited by: Behram N. Kursunoglu, Stephan Mintz and Arnold Perlmutter
Plenum Press, 1996.

Neutrino Mass, Dark Matter, Gravitational Waves, Condensation of Atoms and Monopoles, Light Cone Quantization
Edited by: Behram N. Kursunoglu, Stephan Mintz and Arnold Perlmutter
Plenum Press, 1996.

Technology for the Global Economic, Environmental Survival and Prosperity
Edited by: Behram N. Kursunoglu, Stephan Mintz and Arnold Perlmutter
Plenum Press, 1997.

25th Coral Gables Conference on High Energy Physics and Cosmology
Edited by: Behram N. Kursunoglu, Stephan Mintz and Arnold Perlmutter
Plenum Press, 1997.

Environment and Nuclear Energy
Edited by: Behram N. Kursunoglu, Stephan Mintz and Arnold Perlmutter
Plenum Press, 1998.

Physics of Mass
Edited by: Behram N. Kursunoglu, Stephan L. Mintz and Arnold Perlmutter
Kluwer Academic / Plenum Publishers, Inc., New York, 1998

Confluence of Cosmology, Massive Neutrinos, Elementary Particles, and Gravitation
Edited by: Behram N. Kursunoglu, Stephan Mintz and Arnold Perlmutter
Plenum Press, 1999.

Preparing the Ground for Renewal of Nuclear Power
Edited by: Behram N. Kursunoglu, Stephan Mintz and Arnold Perlmutter
Plenum Press, 1999.

Contributing Co-Sponsors of the Global Foundation Conferences

National Science Foundation
Alpha Omega Research Foundation, Palm Beach, Florida

Précis

This conference will focus on the replacement of existing nuclear power plants when they will reach the end of their useful life. The decision to build new plants will have to be taken for the majority of existing plants during the next twenty years.

The questions we will discuss are:

- *Will the nuclear plants be competitive at this time with other sources of energy?*
- *Will the public acceptance of nuclear power be sufficient to allow the construction of nuclear plants?*
- *Will the decision makers and the public take into account the positive aspects for the environment of nuclear energy (CO_2 issue,...) when making the decision of the replacement of present plants?*
- *We will try to discuss the above mentioned points in different geographical contexts which means country by country or region by region.*

CONFERENCE PROGRAM

7:45 AM Registration

8:15 AM **OPENING SESSION**

J. COUTURE (Trustee of the Global Foundation Inc.) Welcoming Participants
B.N. KURSUNOGLU (Chairman of The Global Foundation Inc.) Opening Remarks
C. STARR (Founder and President Emeritus of Electric Power Research Institute)
"The Future of Nuclear Power and How to Get There"
M. DOUCET (Secretary General, WEC)

10:00 AM Coffee Break

10:30 AM **C.PIERRET** (Secretary for Industry) (Keynote Speaker)

10:45 AM **SESSION I: NUCLEAR COMPETITIVENESS: THE ECONOMICS AND POLITICS OF NUCLEAR ENERGY**

Moderators : **J.D. LEVI** (Senior Vice President of Framatome) and
S.T. BREWER (former President of ABB Combustion Engineering)

Dissertators: **SHELBY BREWER**
"Nuclear Business and Finance, in the context of massive restructuring of the electric utility sector in the US"
P.SAVELLI (Deputy Director General, Nuclear Energy Agency)
"Nuclear Competitiveness Worldwide"
H.MAEDA (Executive Managing Director, Kansai Electric Power Company)
"Future Competitiveness of Nuclear Electricity in Japan"
F.BOUTEILLE (Executive Vice President, Nuclear Power International)
"The French and German Project of the European Pressurized water Reactor (EPR)".

Annotators: **J.PAFFENBARGER** (Electricity expert-International Energy Agency) who will present a contribution on the position of the IEA on electricity market competition and nuclear power

Session Organizers: **VIEILLARD BARON** (Director of Framatome)
SHELBY BREWER

12:30PM Lunch Break: Luncheon Speaker: **C.MANDIL** (General Director of Energy and Raw Materials, French Ministry of Industry).

2:30 PM **SESSION II: NUCLEAR COMPETITIVENESS: NUCLEAR POWER PLANT EVOLUTIONS**

Moderators: **S. T. BREWER** and **B.VIEILLARD BARON**

Dissertators: **P. LECOCQ** (Director ,EDF)
"EDF Actions to Improve the Competitiveness of Nuclear Energy in the Future"
G. DAVIS (Director, ABB Combustion Engineering)
"Competitiveness of Nuclear Plants in a deregulated market place"

* Nobel Laureate

B.BARRE (Vice President, Commissariat à l'Energie Atomique)
"Contribution of Research and Development to the Nuclear Power Competitiveness"

R.W.GALE (Chairman and CEO, Washington International Energy Group)
"Why Nuclear is Competitive and how it can stage a comeback?"

Annotators: **H.FRESLON** (Vice President, Nuclear Operations, Framatome)

Session Organizers: **VIEILLARD BARON** and **SHELBY BREWER**

4:15 PM Coffee Break

4:45 PM **SESSION III SOCIAL ACCEPTANCE: COMMUNICATION, INFORMATION**

Moderators: **J. PANOSSIAN** (executive Director of SFEN) and **Y. GAGARINSKI** (Deputy Director of DAF and KTG)

Dissertators: **P. HAUG** (Executive Director of DAF and KTG)
"Working towards Public Acceptance in Germany: Past Achievements and Present Challenges"
S. BJURSTROM (Director - S.K.B. Sweden)
"Nuclear Waste and Public Opinion"

J.P. CHAUSSADE (Senior Advisor of EDF)
"Nuclear Communication: How to misunderstand the Wants and Needs Of Public Opinion"
A. HOWARD (Senior Vice President of NEI , Washington DC)
"Public and Policy Maker Support for Nuclear Energy"

Annotators: **J. DE LA FERTE** (External Relations, NEA-OCDE), **D. KYD** (Head of press service of IAEA)
EARLE MARIE HANSON (Los Alamos National Laboratory)

Session Organizers: **J. PANOSSIAN** and **CHARLES YULISH** (Vice President, US Enrichment Corporation)

6:15 PM Conference Adjourns for the Day.

OFFICIAL DINNER

France Amerique

9/11, Avenue Franklin Roosevelt 75008, Paris

FRIDAY, OCTOBER 23, 1998

8:30AM **SESSION IV: SOCIAL ACCEPTANCE: REGARDING WASTE, SAFETY, AND PROLIFERATION**

Moderators: **G. CLARK** (Secretary General of Uranium Institute) and
B. PELLAUD (Deputy General Director of AIEA)

Dissertators: **ROBERT WILSON,** Harvard University
"Safety and Regulatory Issues"
J.PILAT (Visiting Research Fellow, IISS, London)
"Proliferation of Nuclear Weapons"
B.PELLAUD
"With Good Safeguards the Civil Nuclear Industry does not contribute to Nuclear Proliferation"
EDWARD ARTHUR (Deputy Director of Civilian Applications, Los Alamos National Laboratory)
"Plutonium and the Nuclear Future"

Annotators: **MYRON KRATZER,** Washington DC
ROBERT KRAKOWSKI, Los Alamos National Laboratory

Session organizers: **G. CLARKE and Ed ARTHUR**

10:15AM Coffee Break

10:45AM **SESSION V: DURABILITY: RESOURCES**

Moderators: **J.P. ROUGEAU** (Senior Vice President of Cogema) And **E. ARTHUR**

Dissertators: **R.KRAKOWSKI** (Senior Staff Member, Los Alamos National Laboratory) **L. BENNET** (Former AIEA) and **E. BERTEL** (Principal Administrator, NEA)
"Nuclear Fission for Safe, Globally Sustainable, Proliferation-Resistant, and Cost- Effective Energy"

WILLIAM MARTIN (Chairman of Washington Policy and Analysis, Inc)
"Short Term Security of Resources"

C.D.GÉRÉ (Scientific Director of the Defense Studies Foundation, Paris)
"Strategic Aspects of Energy Resources"

P. BECK (Royal Institute for International Affaires)

"The Global Energy Situation in the next Century and its Implications on Nuclear Power"

Session Organizers: **JEAN-PIERRE ROUGEAU** and **ANTHONY FAVALE** (Deputy Director Energy Systems of Northrop Grumman Aerospace)

12:30PM Lunch Break: Luncheon Speaker, **C.GOBERT** (Executive Vice President, Cogema)

2:30PM **SESSION VI: DURABILITY: EFFECTS ON THE ENVIRONMENT**

Moderators: **A. FAVALE** and **J.P. LAURENT**† (Vice President for environment of Cogema)

Dissertators: **K.LEYDON** (Unit Head, Directorate 17, European Commission)
"Energy Futures in a Sustainable World"

S. SANDKLEF (President and CEO of Vattenfall Braensle, Sweden)
"Nuclear Power and Sustainable Development"

J.VAN GEEL (Director, ITE, Kalsruhe)
"Innovative Fuel Cycles"

S.MACHI (Deputy Director General, IAEA)
"Application of Radiation and Isotopes for Sustainable Development"

Annotator: **S.LATEK** (National Atomic Energy Agency, Poland)

Session Organizers: **J.P. ROUGEAU** and **A. FAVALE**

4:00PM Coffee Break

4:30PM **CONCLUDING PANEL DISCUSSION**

"PREPARING THE GROUND FOR RENEWAL OF NUCLEAR POWER"

Moderator: **R.LINKOHR** (European MP, President of the European Energy Foundation)

Annotator: **C.PIERRE ZALESKI** (Deputy Director, CGEMP)

Panel Members: **R.JEFFREY** (President, British Energy), **W.MARTIN** (Former US Deputy of Energy), **F.ROUSSELY** (Chairman, EDF), **D.VIGNON** (President and CEO, Framatome), **S.T. BREWER**
Secretary

6:30PM **1998 Global Foundation Energy Conference Adjourns**

7:30PM Dinner Meeting of the Global Foundation Board of Trustees and Advisory Board

EUROPEAN ORGANIZATION COMMITTEE
GLOBAL FOUNDATION 1998

Mr Jean Couture Trustee of Global Foundation and Chairman of European Sub-Committee	Dr. C. Pierre Zaleski Deputy Director, CGEMP Principal Program Coordinator	Dr. Gerald Clark Secretary General The Uranium Institute Member
Ms. Thérèse de Mazancourt Member	Mr. Jacques Panossian Executive Director of SFEN Member	Mr. Jean-Pierre Rougeau Senior Vice-president of COGEMA Member
Mr. Bertrand Viellard-Baron Director of Framatome Member	Mr. Yasmine Benamour Université Paris Dauphine Secretary	Ms. Nicole Guillard Université Paris Dauphine Secretary

Contents

Chapter III
Durability: Resources and Environment

Chapter IV
Concluding Panel Summary of the Conference

THE FUTURE OF NUCLEAR POWER

Behram N. Kursunoglu

The subject of nuclear power has five fundamental components:

1. Nuclear power carries the highest density of energy and is available.
2. It is competitive with other sources of energy when the reactor power exceeds 1,000 MW.
3. Nuclear reactor malfunction can result in serious accidents.
4. It provides a basis for the proliferation of nuclear weapons.
5. The problem of localized radioactive waste remains a challenge.

The nuclear waste problem should be compared to the problem of fossil fuel waste which is emitted worldwide into the atmosphere in the form of greenhouse gases and has become irreversibly high, affecting the climatic condition of the world. Continuous increase accumulation of these gases in the atmosphere will eventually result in a lower standard of living. There are many ideas with regard to the solution to nuclear waste problems. They include sending waste into space and placing it in a very high orbit around the earth; preservation of waste underground where a lot of salt is found (implying the absence of water); reprocessing of waste to remove very high levels of radioactivity which can be mixed with glass in a process called vitrification as used by the French; etc.

The best fuel for the production of energy is, of course, sunlight. But at present, it still falls short of competing with nuclear or fossil fuel generated energy. In the case of nuclear energy, we need global usage of it in order to produce it cheaply enough and also to throw over it global supervision to prevent proliferation of weapons of mass destruction.

In the past two decades, there has been no new nuclear power plant construction in the United States and therefore, the use of oil and natural gas did not decrease. The current reserves of fossil fuels include one trillion barrels of oil and four thousand trillion cubic feet of natural gas. These numbers do, of course, change on an annual basis. They appear to be on the side of increasing reserves.

There are also coal reserves, mostly located in China, Russia, and the United States. The total energy content is about equal to the energy content of the combined oil and natural gas reserves. The use of coal is known to be the source of copious emissions of greenhouse gases, much more so than oil and gas.

Anticipated political gains from the transfer of fossil fuels exceed their economic and environmental consequences. For example, the Middle East is no longer the sole producer of crude petroleum. The new situation includes the participation of the Caspian region.

Preparing the Ground for Renewal of Nuclear Power, edited by
Kursunoglu *et al.* Kluwer Academic / Plenum Publishers, New York 1999.

Approximately half a trillion barrels of oil reserves are available from the Caspian region and need to be transported to the world markets, mostly via pipeline. In this case, the oil from Kazakhstan, Turkmenistan, and Azerbaijan is expected to flow from these countries through the Baku-Ceyhan pipeline. All these sources and the transportation of the oil impact the use of nuclear power.

We have two alternatives. One is that we use all these fossil fuels and exhaust them within approximately one to two hundred years, meanwhile suffering the environmental consequences. The second alternative is non-use of these fuels and a shift in emphasis to nuclear power, accepting it with all its virtues and vices, still coming out better off than the first alternative. Let us have a brief survey of fossil fuel produced energy.

Table 1. Oil and Natural Gas Production by Region

REGION	OIL (10^{12} BARRELS)	NATURAL GAS ($4x10^{15}$ CUBIC FEET)
Middle East	65.2%	30.7%
Latin America	12.5%	5.8%
USSR & Eastern Europe	5.9%	38.3%
Africa	5.9%	6.7%
Asia	4.5%	7.1%
North America	4.2%	6.5%
Western Europe	1.8%	4.9%

OPEC provides 39% of the world's energy, which could keep the global economy energized for many decades. As a mobile fuel for transportation, oil cannot be replaced by anything in the short run. However, using natural gas as a fuel for fleets of cars and trucks could displace at least 10% of eight million barrels of oil a day (about equal to total imports). Furthermore, Soviet oil (3.2 million barrels produced per day) could and would play an important role (just as Soviet gas did when pipelined to Western Europe over the past six years) if the West extends the needed know-how and technology to the Soviets to improve production.

The total gross world production (GWP) at the current economic standing

Table 2. Total Gross World Production (GWP)

COUNTRY	PRODUCT
United States	\$ 5.5×10^{12}
European Union	\$ 5.0×10^{12}
Japan	\$ 3.0×10^{12}
Other	\$ 11.5×10^{12}

of \$25 trillion is, to a great extent, being energized from the Persian Gulf region which includes, besides Iran, the United Arab Emirates, Libya, and Nigeria. The non-OPEC countries, such as Britain, Norway, and the Soviet Union play a smaller role in the energy balance. The supply and demand difference (i.e., S-D/S=?%), if negative, raises the price of oil. It is a fact that the "spark plug" of the economic machine will not spark without oil. Thus, OPEC, by itself, constitutes a "superpower" in global economics.

The current world need from OPEC is around 24.2 million barrels a day. Despite the Saudi oil production jump from 5.4 million barrels at the beginning of August to 8.2 million barrels at present, OPEC oil production reach only the level 23.2 million barrels a day. Thus, the negative supply and demand difference, coupled with the fear of supply shortage (war), leads to high market prices. But it must be noted that the price increases have put pressure on consumption, reducing it by a million barrels a day. The current world demand for oil per day stands around 65 million barrels, 12 million barrels of which are consumed by the USSR, China and Eastern European. Saudi Arabia has the biggest worldwide spare oil production capacity right now and is also the only country that can go further.

The recent Persian Gulf crisis may be laying the foundations for the end of the age of oil not only politically and economically, but there are environmental reasons for the coming gradual sinking of the star of oil. It may also be the end of oil sheiks, emirs, etc., and perhaps the end of dictators. In case of another Persian Gulf war, the market price of oil could vary from $40 to $80 per barrel and, if the cost of the war is included, it could go higher (e.g., $150 a barrel) compared to the current $10 per barrel oil price. If the price increased it would put a squeeze on refining, marketing, and petrochemicals. In this case, self-imposed conservation could reduce consumption by 50% (a figure that has been experientially observed), resulting in the daily use of 30×10^6 barrels worldwide. With the daily cost equaling $150 \times 30 \times 10^6 = \$4.5 \times 10^9$, which is even at the greatly reduced rate of oil consumption, the cost is still 35% higher than the pre-Gulf war price. The daily barrage of warnings, signals, etc., to Saddam Hussein are numerous enough to form a wall around him. Any new warnings are not received, they simply bounce off.

We must move with the new political and economic implications of NATO. World leaders have included the Russian Federation in these efforts, along with Japan and other powers, thereby designing a new world order where a gradually emerging BOPEC is abandoning the dependence on Persian Gulf oil and letting its price fall freely to make it possible for "developing countries" to benefit. Industrialized countries should gradually switch to the use of alternative energy sources, such as non-OPEC oil (except Venezuela), natural gas, other mobile fuels and, eventually, evolving to a world economy energized entirely by electricity. We must develop electric cars to a technological level that assumes the role of oil at 100% efficiency.

The use of nuclear energy (despite Chernobyl and Three-Mile island) should be made possible and acceptable by reeducating the public. The signature of the nuclear era (the swiftly ascending mushroom cloud) has left an eerie fascination and fear in the minds of everyone who observed it for the first time and every time thereafter. With this mushroom cloud, the world entered, presumably forever, a new political, economic, and social order. The blasts at Hiroshima and Nagasaki not only announced the awesome arrival of the nuclear age, they also detrimentally affected public perception of the word "nuclear" and public acceptability of the peaceful uses of nuclear energy. Even the term "nuclear magnetic resonance," an observed event in nuclear physics, has been assigned the nomenclature MRI (magnetic resonance imaging) in its radiological applications in medicine to avoid the use of the word "nuclear."

However, no argument has yet been presented to exclude the role of nuclear power in meeting the energy needs of all nations in the future. When will there be a complete acceptance of nuclear power? I do not know. But it will happen.

THE 21st CENTURY'S INTERNATIONAL NEED FOR NUCLEAR ENERGY

By Bertram Wolfe

NUCLEAR POWER TODAY

In the industrialized world today there is generally a very low growth rate of energy use. Further, the discovery of large low cost gas supplies in the past decades have made gas fired electrical generating plants a simple, economic, means of supplying new capacity. As a result, the expanded use of nuclear power in the industrial countries has essentially come to an end. Indeed, in Sweden, Italy, and now in Germany, there are legal impediments to the use of nuclear energy; and in the U.S. there has not been a new nuclear plant order (which was not canceled) since the Arab oil boycott of 1973. On the other hand, in the developing countries, primarily in Asia, the need for added electrical supplies is growing, and the gas and fossil fuel supplies in general are limited, so that expansions of nuclear energy are taking place and further expansions are contemplated.

The point is that *need* has been the driving force for the use of nuclear energy, and need has decreased in the industrial nations. Before the Arab oil boycott of 1973 energy use, and nuclear energy, was in a state of major expansion in the U.S.. But after the boycott, because of increased energy costs, US energy growth greatly diminished, and this resulted in a US surplus of electrical capacity. The nuclear energy need, and growth, ended. On the other hand, after the boycott, because of concern over the cost and risk of dependence on imported energy, France greatly expanded its nuclear generation capacity, as did Japan and other nations. They felt a *need* for nuclear energy. But today, in Europe and the US there is substantial energy available and nuclear energy is at a standstill.

Preparing the Ground for Renewal of Nuclear Power, edited by
Kursunoglu *et al.* Kluwer Academic / Plenum Publishers, New York 1999.

The key point of this paper is that the situation will greatly change in the future; and there will be a worldwide *need* for a major expansion of nuclear energy. The question is whether the industrial nations will recognize the coming need, and take appropriate, timely, action to prevent unnecessary world hardships. Or will delays continue until the hardships arrive; and the solutions are more difficult.

THE COMING NUCLEAR NEED

It is projected that by the middle of the next century the undeveloped world population will double from four to eight billion people and the total world population will reach ten billion. If the third world increases its standard of living, and world energy use stabilizes at a third of the per capita use in the U.S. today, then world energy requirements will triple, compared to the present world use. Such a situation poses immense energy problems. The world faces potential environmental disasters from fossil fuel burning and potential international hostilities over scarce fossil fuel supplies. A further problem for our descendants is the early depletion of fossil supplies. There appear no solutions to these problems other than a major expansion of nuclear energy.

If one assumes that nuclear energy will be needed to provide a third of the world's energy by the middle of the next century, then present nuclear capacity would have to increase by a factor of about fifteen, from about 350 GWe to some 5000 GWe of nuclear capacity; or to some 8000 GWe if nuclear power were to provide half of the world's needed energy. This would require the addition of a hundred to two hundred nuclear plants per year over the next fifty years; an order of magnitude increase in the production rates of the past few decades.

Can this be accomplished? The answer is yes, from a physical and technical standpoint. But what is likely to be required is world recognition of the need, and an integrated world program.

SAVING THE FUTURE WITH NUCLEAR ENERGY

One might note that in the late sixties and early seventies some thirty to forty nuclear power plants were being ordered each year in the U.S.. It was expected that there would be a thousand plants in operation by the end of the century. There seemed little question that this could be accomplished. However, the decrease in energy growth in the U.S. after the 1976 Arab oil boycott, caused a surplus of electrical generation. In the past two decades U.S. nuclear plants increased in number by only some 70 plants rather than the 1,000 odd plants previously contemplated. Further, in the U.S. and most of the rest of the industrialized world the slowdown in nuclear energy growth has led to loss of nuclear equipment manufacturing capability. For instance, there are now no nuclear pressure vessel manufacturing facilities in the U.S., and throughout the world there has been a decrease in the facilities which could mass produce nuclear components to the quality required. Thus to meet future world needs, quality nuclear component manufacturing capability must be greatly expanded.

Over the past decades the key nuclear design problems have been identified, and have (hopefully) been resolved in the new light water and heavy water nuclear plant designs. In the U.S. the new designs have received what are referred to as "standardized" licenses which allow plants to be built expeditiously as long as they retain the design characteristics and requirements for which the license was issued. This is intended to eliminate the long 10 to 20 year licensing delays of the past that resulted from each plant being considered a new design process. This new US licensing process is not very different than the licensing procedures in France, where standardization of nuclear plant design has permitted the construction of some 55 new plants on six year timetables in the past two decades. They now provide over two thirds of French electricity.

One should also note the improvement in nuclear plant operating efficiencies that are continuing to take place. For example, refueling periods which have taken as much as three to six months of down time, have in many plants now been reduced to less than thirty days; and operating staffs have been significantly reduced to decrease operating costs. In both of these cases the operating quality has actually improved.

Thus, one should recognize that although the present nuclear plant designs have had their problems, these have been attended to, and there are now designs and procedures that eliminate past problems, and allow plants to operate reliably for decades.

Despite the improved reliable designs, the low level of demand means that the industrial countries can continue without nuclear expansions. But if one projects, as above, the need for many tens or a hundred of new nuclear plants each year for the coming decades then governments and the nuclear industry on an international basis should now plan and prepare for the needed expansion.

INTERNATIONAL ACTIONS FOR THE FUTURE

The following actions should be considered:

- Nuclear plant licensing should be contemplated on an international basis with the ABWR, the System 80+, the AP600, French-German design, the Candu and the upgraded VVWR designs reviewed and hopefully accepted so that such plants (and improved versions) could be built as needed throughout the world without major licensing delays.

- Plans should be developed now for expansion of manufacturing capability. Such expansion should be done in a manner to assure that the expanding nuclear component needs can be met economically, meeting quality standards, and on schedule.

- Plans should be developed, and approved internationally, for the organization of plant operating staffs, for their training; and for plant operating procedures.

- International agreements should be expanded so that an organization such as the IAEA has the authority to perform inspections to confirm that safety procedures and requirements are being met, and to assure that there is no diversion of materials that could be used in nuclear weapons.

- Plans should be developed to permit easy communication nationally and internationally between nuclear plants so that positive experience and procedures, and effective modifications can be shared between them. Organizations such as INPO, WANO, and COG could be expanded and/or combined to meet this objective.

- The accommodation of nuclear wastes should be examined on an international basis. This is especially so, since the expansion of nuclear power is apt to require the future use of the fast reactor which will use the plutonium in the spent fuel. International reprocessing facilities would seem to make good sense.

- The development of the fast reactor should proceed now so that it will be available when fuel for thermal reactors becomes less plentiful and too costly. The fast reactor can provide an essentially unlimited world energy supply.

- The industrial nations should examine the financial needs of the developing countries and seek means to provide the financing of needed nuclear plants, which will reduce the use of greenhouse emitting fossil fueled plants, and possible hostilities over short energy supplies.

CONCLUSIONS

In summary, the world is facing a major energy problem that could have disastrous effects if it is not solved. The major expansion of Nuclear Energy internationally may be the only means of reaching a solution. The past experience, and new nuclear plant development by the nuclear industry, places nuclear energy in a position to meet future world needs.

In the near term, nations can ignore the coming energy problems. But, in view of the potential world tragedies that could be forthcoming, delays should not be tolerated. We should move on an international basis to help provide a safe, and satisfying energy future for the coming decades.

Dr. Bertram Wolfe retired in 1992 as a V.P. of General Electric, and head of its peaceful nuclear energy organization He has since been an independent consultant, on several Boards of Directors, and on a number of advisory committees. He is a Fellow and past President of the American Nuclear Society; a member of the National Academy of Engineering, and a Professional Engineer in California. He has received a number of industry honors, and is the author of over a hundred publications on Energy and Nuclear Energy.

[illegible] The [illegible] should examine the financial needs of the developing countries and seek [illegible]

means to provide for financing of needed nuclear plants, which will reduce the use of [illegible]

[illegible] possible [illegible]

[illegible]

Observations on The Future of Nuclear Power and How to Get There *

Chauncey Starr

President Emeritus
Electric Power Research Institute (EPRI)

I am pleased to have been given this opportunity to address this International Conference on Preparing the Ground for Renewal of Nuclear Power by video, although I regret missing personal discussions with each of you. As some of you know, I have been associated with nuclear power development since its earliest 1940's days. Viewed from this half-century of experience, I will summarize my perceptions on the future of nuclear power.

The provocative title of this meeting implies a dormancy for nuclear power which may only be partially descriptive of the real situation. In many industrial countries nuclear power has been politically fenced in (corralled, in US slang) by antinuclear quarantines politically building on public fears. The inadequate performance of some nuclear utilities has increased such fears. This is a regrettable outcome to those of us who have been dedicated to making nuclear power a nonfossil energy source for the coming millennia. However, some optimism is justified by slowly developing trends world-wide.

Most importantly, it is inevitable that in the coming century the global need for nuclear power as a major electricity source will be publicly accepted. The driving forces of global population growth and economic improvement will create an energy demand so large that fossil fuel resources will be strained. I have discussed the quantitative aspects of these matters elsewhere (1), but only a few numbers are needed to illustrate the issue. **Demographic projections conservatively estimate that in the next half century global population will go from 6 billion to 9, and probably reach 12 billion by year 2100.** This population doubling in 100 years is modest compared to the 4x global population growth of this past century.

Simultaneously, per capita energy consumption will rise with economic growth, with a wide range of projections. As an illustration, consider that the US per capita energy consumption is now more than 10x greater than that of the less developed fraction of the world's population, and about 5x the global average. As a guesstimate, assume that sometime before the end of the coming century, the global average per capita consumption reaches half of US today, i.e.

Preparing the Ground for Renewal of Nuclear Power, edited by
Kursunoglu *et al.* Kluwer Academic / Plenum Publishers, New York 1999.

goes from 20% to 50% of US today (about the same as Eastern Europe and Former Soviet Union today). **Combining the above increases (population 2x and per capita use 2.5x) results in a 5x increase in global annual energy consumption.** Of course these are speculative projections, but compared to past trends they are reasonable, perhaps even conservative as they embody a modest global economic growth.

The immediate question is how such energy needs would be supplied. A study of technical trends provides a judgmental projection of the year 2060 mix of alternative energy sources (1), as shown in **Chart I**. This projection assumed economic growth of 2.3%/yr (a 30yr doubling time) and very optimistic 40% reduction in demand through efficiency and conservation. Global economic growth rate is the most uncertain parameter. In this 2060 projection, fossil fuels supply about 60% of the total annual energy use (as compared with 85% today), with annual global carbon emission about 1.8x that of today. Although nuclear capacity is projected to grow in this scenario to about 10x today's level, it will only contribute 17% of total energy (compared with 7% today). Hydro and renewables are very optimistically assumed to share the 23% remaining (compared with 8% today of mostly hydro).

World Net Electricity Consumption

1000 Terawatt hrs: (10^{12} kWh)

Region	1997	Projections 2020	Projections 2060
Industrialized nations	7.5	11.3	28.0
United States	(3.2)	(4.3)	
EE/FSU	1.5	2.2	
Developing countries	3.4	9.6	19.0
Asia	(2.1)	(6.7)	
China	(1.0)	(3.6)	
Other Asia	(1.1)	(3.1)	
World Total	12.4	23.1	47.0

If in the next few decades it is clearly established that carbon emissions are a serious global threat, nuclear power growth will become urgent. It is certainly hypocritical and deceitful for environmentalists to purposely vilify and deprecate the importance of the global insurance provided by the nuclear option.

The real dilemma for environmentalists who urge massive reduction of energy use is balancing a far-future global consequence of an uncertain climate change versus the known consequence today of world-wide regional poverty. In a democratic world, public acceptance of mandated sacrifice of the benefits of economic growth is not likely. Even in the relatively affluent US, the best health services, education, and care of the aged, are not fully available to all. Further, these are moving targets that are always ahead of us.

Thus, in industrial countries today, and particularly in the US, an essential global warming precaution is to continue the long-term development of the nuclear power option and closure of the back end of its fuel cycle, including the options of spent fuel storage, reprocessing and fast reactors. These options should be functional when we need them. Obviously, political support for such engineering demonstrations should be underway now in advance of the foreseeable need.

So much for energy futures; now let us turn to the electricity component. Electricity grows almost twice as fast as total energy. Chart II shows that a 4x increase in electricity generation by the middle of the coming century is likely, even with maximum conservation and efficient use. This will impose a stupendous annual fuel demand on fossil fuel resources. Although this scenario cannot anticipate all the significant changes in the coming decades, it is a conservative basis for future strategies. **It is the magnitude, not the accuracy, of these estimates that is important. Also, the time scale may be much shorter if the aspirations of political leaders for economic growth in developing countries are fulfilled.**

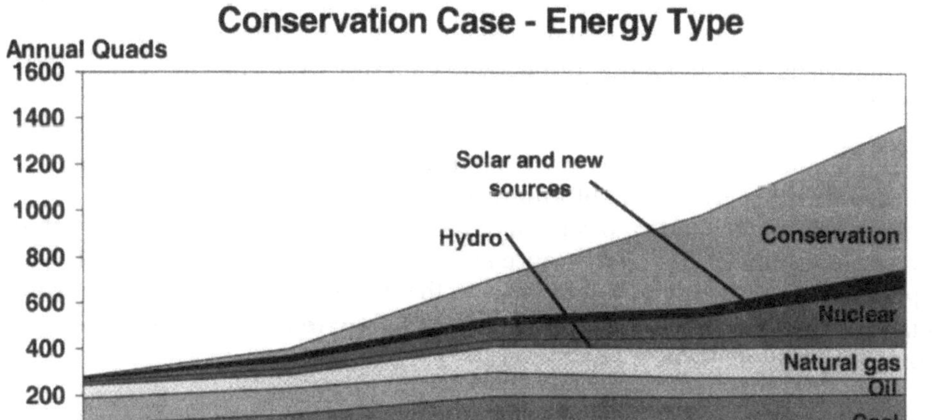

So, the high probability of increased costs of fossil fuels to meet the foreseeable large multiplication of demand defines the dominant role of nonfossil electricity alternatives, and particularly of nuclear power in a future mix (2). The strategic policy question is what will this mix be, when will it be needed, and how paid for. Nuclear and hydro are today the only nonfossil sources that offer reliable base load kwhrs on a substantial scale. So countries that depend on imported fossil fuels, and have limited hydro, must eventually turn to nuclear to ensure security of an economic electricity supply.

Our past experience suggests that it takes almost a half-century for a new electricity source to become a major contributor. So, today's ample availability of low-cost fossil fuels provides the time window to gain experience with nonfossil systems, to determine their relative merits, to establish their supporting systems, and to find the special niches where they fit best. The natural limitations faced by renewables are well known. Both solar and wind are intermittent, and biomass and geothermal are site specific. They are mostly niche additions to conventional systems and generally survive on large subsidies. Total dependence on them for reliable base load power is theoretically possible only with large costly storage facilities (pumped hydro, compressed air, etc.), and thus at system costs at least ten or more times greater than today's costs. The growth of the capital resources of the world could not conceivably be adequate for such extreme investments.

Nevertheless, such total dependence on renewables (solar, wind, biomass, geothermal, ocean waves, tides, etc.) is the visionary goal of today's antinuclear environmental community. That is an anti-social doctrine that can only create havoc to global welfare if seriously implemented. It is unfortunate that some politicians in industrial societies have embraced this seductive mirage as a popular solution to the future energy needs of the real world. They either suffer from the childlike innocence of the ignorant, or perhaps even worse, are knowingly engaged in political duplicity to gain support from a those worried about nuclear and from committed environmentalists.

Eventually, political priorities must balance the practical energy trilemma of population, prosperity, and pollution. I believe that these political balances that operate today to constrain the nuclear option must eventually shift to create its ultimate global acceptance and expansion. **So I expect a global supply push for nuclear power to grow in the next several decades, in spite of today's mixed public acceptance. We should prepare now for that more promising future. How?**

The timing of this process depends on two major factors; public trust in the integrity and competence of the nuclear industry; and reasonable public expectation that the nuclear option is a better balanced choice than the alternatives. The industry has strong economic incentives to operate safely and develop public trust. We all hope it does so. Public acceptance is a more difficult target.

Based on my own experiences in public debates and discussions, I suggest a few basic principles for reinforcing public confidence. The public has more faith in demonstrated performance by utilities than in government regulatory agencies. The public distrusts concealment of significant information or events. Organizational transparency is the key to integrity. The public is skeptical of self-applauding speeches and advertisements. The public is shaken by signs of managerial incompetence or carelessness, especially if nothing is done to remedy such situations. And the public is not interested in becoming scientifically informed, or in the details of government regulations. But the public is always interested in their exposure to personal risks (health and costs) from any energy source.

Measured against these simple "human relations" criteria, our nuclear industry has not done well. It is worth noting that France's EDF has had a reasonably successful program. Most of us know of other successes in getting public support, but it takes only one major fiasco to quickly wipe out years of confidence building. Further, we face the persistent attack of the antinuclear movement, a well organized and funded minority, which has now worked its doctrine into the political structure and news media of many countries. The nuclear industry has not created a credible counter-campaign. **Even harmless low-level radiation, naturally always with us, and also an unavoidable byproduct of all nuclear activity, has been imaged as a public evil by the antinuclears. The uncommitted public majority must be confused. It is obvious that the fear of radiation, even at the lowest levels, is one of the major obstacles to public acceptance of nuclear power. Amazingly, the nuclear utilities have not supported the efforts of that small group of scientists actively challenging the regulatory concept that radiation at any level is dangerous (3). It is important that we all give support to them.**

Public choices among energy alternatives are usually based on familiarity and superficial impressions of their merits. To develop public awareness of nuclear's merits, we should undertake and publicize a comparative benefit/cost/risk study of all alternatives, including their intangible long-term social costs of health and safety. A comparison of alternatives with nuclear has been discouraged by electric utilities as their non-nuclear electricity sources which have rarely been subject to public attention. However, now the environmentalists are attacking fossil fuel and hydro sources. Some have seriously proposed that CO_2 be a regulated pollutant. Of course, nuclear remains a special target. **A comparative benefit/cost/risk study is an immediately feasible task, and we should organize to do this now if we believe in the merits of nuclear power.**

As a crucial longer-term issue, we must address the antinuclear teaching in the schools of many industrial countries where antitechnology and environmentalism are now the politically correct posture. This is a real concern in the US, where even antiscience surfaces. A joint program of all national technical and scientific societies is needed to work with teacher's groups to educate them, and the media, on the role of technology, energy, and radiation, in sustaining people's welfare, particularly health and safety. Such a joint program requires persistent and organized effort, but it is essential if nuclear power is to be preserved for the next generations.

So, recognizing the uncertainties of all global scenarios, I remain optimistic about nuclear power's future, and of the societal importance of our continued support for it. Thanks again for the opportunity to talk to you.

*Presented at the Global Foundation "International Conference On Preparing The Ground For Renewal Of Nuclear Power", Oct. 22-23, 1998, Paris.

1. "Energy Sources: A Realistic Outlook", Starr, Searl, and Alpert, Science v.256, pp. 981-987 , 15 May 1992.

2. "The Next Oil Crises Looms Large – and Perhaps Close", R. A. Kerr. Science v.281, pp. 1128-1131, 21 August 1998.

3. Radiation, Science, and Health, Inc., Box 843, Needham, MA 02194, U.S.A. Phone/Fax 617-449-2214

NUCLEAR POWER AND THE U.S. TRANSITION TO A RESTRUCTURED, COMPETITIVE POWER GENERATION SECTOR

Shelby T. Brewer and Rayburn Hanzlik, Esq.

Managing Directors
Brewer-Hanzlik Nuclear Partners, LLC[1]
McLean, Virginia 22101

INTRODUCTION

U.S. electric utilities are undergoing a major transition from economically regulated and vertically integrated monopolies to structures which allow or mandate competition (and deregulation) of the *power generation* function. The power transmission and distribution functions will remain regulated monopolies.

Wholesale competition was enabled and encouraged by the Energy Policy Act of 1992 and the Public Utilities Regulatory Policy Act of 1978 (PURPA). The next step -- underway now -- is *retail* competition and deregulation, which allows individual residences, factories and other consumers to select their own source of power. Unlike wholesale power deregulation, which came through federal legislation and regulation, retail deregulation is largely in the hands of the states. By the end of 1998, over one-third of the states passed retail electric utility restructuring legislation, and the other two-thirds are involved in a variety of restructuring activities.

These changes will have profound impact on the generation mix. This will prove to be one of the most pervasive paradigm shifts in U.S. economic history. The availability of a reliable and economic electricity supply has been a central component of the modern U.S. economy. With this component now facing the harsh discipline of deregulation and free market forces, we believe that the structural changes will have a positive impact on nuclear power.

[1] Brewer-Hanzlik Nuclear Partners, LLC is an investment holding company building an integrated enterprise in the nuclear industry, with holdings in independent nuclear power plants and companies providing selected services to the nuclear power industry and federal government.

Preparing the Ground for Renewal of Nuclear Power, edited by
Kursunoglu *et al.* Kluwer Academic / Plenum Publishers, New York 1999.

In truth, the forcing of power generation to march toward an Adam Smith economy is an elixir for nuclear power. Fortuitously, this comes at a time when there is also a growing (albeit begrudging) political recognition of the environmental benefits of nuclear power in the face of global warming and the imperative to sustain or raise the standards of living around the world.

We have previously nagged that the nuclear sector must place less emphasis on the elegance of our technology, and waste less time cursing the political darkness and asking for governmental coddling and subsidies; instead, we urged that more attention be focused on nuclear power generation (and its supply infrastructures) as *competitive businesses*. To achieve this, understanding and working in the new utility deregulation/restructuring milieu is fundamental.

Unless old mental habits change, we in the nuclear business will repeat past mistakes and appear to the political and the financial communities as an outmoded and dependent industry without vision and virility. The deregulation/restructuring movement offers immense opportunities and exciting challenges which we should recognize and seize. But we have to be acutely attuned to business and financial realities and the entrepreneurial opportunities, and jettison our technological chauvinism, bred by the fixation on technology within our industry.

The purpose of this paper is to examine the opportunities and challenges for nuclear power in this coming era of competition and structural change. In particular, we advance the idea that an entrepreneurial perspective is necessary and can succeed, and that this *modus operandi* can, in fact, lead to the renewal of nuclear power in the U.S.

EVOLUTION OF U.S. ELECTRIC UTILITY STRUCTURE

Since the beginning of electric utilities, the power supply industry has been regarded as a benign monopoly, or a group of monopolies, with geographic charters. After all, it was (and remains) impractical to have redundant power lines crossing over one another. As for the power generation function (a subset of a utility's responsibility), this, too, has been regarded as appropriately managed by a monopolistic structure for a number of practical reasons: the large investment required; need for economies of scale; requirement for centralized planning; operational efficiencies. Power companies were given charters which allowed them to generate, transmit, and distribute power within geographically defined markets. Since these companies were, in effect, authorized monopolies, the pricing of their product (power) had to be regulated by the government.

What is now underway in the U.S. is the "de-monopolizing" and economic de-regulation of the power generation function of the utility industry. (At this point in time, there are no serious efforts to de-regulate the transmission and distribution functions of the industry.)

Four pieces of U.S. federal legislation are fundamental to the evolution of the legal, financial and institutional structure of the power supply sector:

- Federal Power Act of 1920
- Public Utilities Holding Act of 1935
- Public Utility Regulatory Policies Act of 1978
- Energy Policy Act of 1992

The last two of these laws initiated the movement to restructure the power generation function for wholesale transactions (sales of power to utilities for resale). Retail de-regulation and restructuring (sales by a power generator to end-users) are under the jurisdiction of the states.[2]

Federal Power Act of 1920. This legislation clarified state and federal regulatory (economic) jurisdictions. The Act provided for a Federal Power Commission (now the Federal Energy Regulatory Commission or FERC) to regulate prices of power sold on a wholesale basis between utilities or when interstate sales were involved. Under the Act, prices charged to end-user customers (retail prices) were regulated by the states through public utility commissions (PUC).

Public Utility Holding Company Act of 1935 (PUHCA). PUHCA was enacted to correct several abuses in the utility industry. One such abuse was the use of electricity revenues by some utilities to "cross-subsidize" non-power-related and non-regulated businesses in their portfolios. Procurement abuses were also evident. The PUHCA legislation sought to confine investor-owned utilities to power supply and required that a utility's power supply market be geographically contiguous. A central feature of PUHCA is a provision that limited utility holdings by non-utility companies to 10% of their voting stock. Holdings at or above this threshold risked triggering an SEC requirement that the holding company divest its non-utility interests. For example, if Ford Motor Company's holdings in utilities reached or exceeded 10% of its portfolio, then Ford could be required to divest its automotive businesses.

Public Utility Regulatory Policies Act of 1978 (PURPA). This act was one of the first legislated steps toward breaking up the regulated, vertically-integrated utility monopoly structure. Ironically, it tilted the market playing field in favor of so-called "renewable" (wind, solar, biomass, hydro) and cogeneration energy sources and actually moved the industry a step away from Adam Smith. PURPA *required* utilities to purchase power from power generators using these energy sources and under long-term contracts.

The pricing of PURPA power was based on what was termed the "avoided cost" -- the price that the utility would pay for such power from existing sources. Additionally, such prices escalated under an assumption of rising prices over the contract life. These prices are now above today's market prices, but they are still passed on to the consumer (the amount above market price is termed the "stranded cost"). The Edison Electric Institute estimates that the stranded costs from PURPA power sources will cost consumers $40 billion over the life of their contracts.[3]

A rarefied class of technologies was enshrined by PURPA, and generation plants employing such technologies were termed "Qualified Facilities." QFs included:

- Small (<80 MWe) solar, hydro, wind, biomass units
- Cogenerators (no size limit) with at least 5% of energy output non-electrical and having "industrial useful purpose(s)."

---

[2] The states regulate only investor-owned utilities, not those that are government-owned. Thus, the impact of the restructuring of the power generation function impacts only investor-owned entities.
[3] *The States: Leaders in Electricity Power Industry Restructuring*, August 1998, Edison Electric Institute.

Utility ownership of QFs was allowed, but limited to 49%, and QFs were exempt from the provisions of the PUHCA. Determination of economic indifference (avoided cost method) was left to the state PUCs, which led to a patchwork of arbitrary and complex rules.

PURPA has many flaws, and should have been repealed in the 1992 Energy Policy Act. Besides tilting the power market in favor of avant-garde energy technologies (which have not proven economic or competitive, and which ratepayers have subsidized), PURPA has added billions of dollars of extra costs (from QF stranded costs) to a power generation economy that is learning Adam Smith. And a significant side-effect of PURPA was its opening of the door to market domination by gas-fired electric power plants through the cogeneration provision.

Energy Policy Act of 1992. This act is the most recent fed legislative step in the progression toward a competitive power generation market. The 1992 Act dealt with wholesale power transactions and defined a new facility species, the Exempt Wholesale Generator (EWG). A utility or other entity can build, own and operate an EWG in another utility's service area. An EWG has the following characteristics, as opposed to PURPA's QF:

- It can be based on any technology, fuel, with no size limitations;
- There is no mandatory power purchase requirement as in PURPA;
- EWGs can be built (and owned) by subsidiaries of existing utilities without the 49% ownership limit imposed by PURPA on QFs.

Transmission and distribution access for an EWG was handled by delegating the subject to FERC, which issued Order 888 requiring investor-owned utilities to provide "open access" to their transmission lines to other power generators.

The 1992 Act did not repeal PURPA -- it simply made PURPA largely irrelevant. The only advantage of QF status today is the mandatory purchase provision, but that is a diminishing advantage because the QF cogeneration market is largely saturated. This saturation, in fact, was one of the political drivers for the 1992 Act, as the natural gas industry wanted a new market unconstrained by the PURPA QF cogeneration definition. The 1992 Act, which provided this new market, has been jokingly referred to as "The Natural Gas Bill of 1992."

THE NEXT STEP -- RETAIL DEREGULATION AND RESTRUCTURING

The 1992 Energy Policy Act and PURPA legislation dealt with *wholesale* power sales -- sales between utilities and sales from a non-utility power generator to a utility. The next step, beyond the 1992 legislation, is to inject competition into the *retail* power generation market -- the sale of electricity from a power generator directly to ultimate customers such as homeowners and factories. Visualize homeowners having options other than their local utility company from which to buy electricity. The idea is similar to the competition that came to long distance telephone service in the U.S. Perhaps consumers will soon be hounded -- typically at dinner time -- by marketing calls not only from AT&T and Sprint, but from power generating companies.

Although some federal legislation for retail restructuring has been proposed, none has been enacted or appears imminent. The states are leading the movement to retail competition, with 18 of the 50 states enacting legislation by late 1998 to guide the transition to retail competition. State legislative and regulatory attempts at retail competition vary widely in substance, detail, and schedule, and what ultimately emerges is certain to be a heterogeneous array of constraints and opportunities.

With the movement underway towards a deregulated power generation market, the federal government is busy trying to figure out how to correct the economic imbalances created by its previous actions. The billions of dollars of stranded costs, imposed on the market by PURPA and by the federal micro-regulation of the nuclear industry, hang over a competitive power generation market like a massive debt and are an invitation for federal action. (We should note from experience that inviting the federal government to enact over-arching homogenizing legislation and regulations, however well-meaning, always brings unintended side-effects and usually harmful results.)

There are several proposals in Congress to deal with stranded costs in the face of deregulated power generation. One proposal for national legislation would require stranded cost recovery that favor both nuclear power and the renewables engendered under PURPA. The resultant costs to consumers would be in the tens of billions and postpone true market-based pricing for decades. At the opposite extreme, another proposal would prohibit any stranded cost recovery. This would not only injure utility shareholders, but would terminally harm nuclear power. This proposal is jokingly referred to as the "Natural Gas Act III." Our point is: federal intervention to address this imbalance will surely prove as flawed as other federal forays into the energy market. May Adam Smith rather than Uncle Sam work his magic here.

THE STATE OF CORNUCOPIA

To illustrate the process of deregulation, we have created a mythical state called the State of Cornucopia, which is an attempted composite of a typical state and how it would deal with the deregulation of power generation. The purpose is to focus on some of the fundamental steps and patterns seen in this process.

Cornucopia has passed legislation to mandate and enable retail competition in electrical power generation. The PUC is implementing this legislation with codes, rules, and procedures. The Cornucopia PUC, by the way, has seized this as a new opportunity to expand its bureaucratic mission, and has published on its Web site that its scope and budget will increase in order to manage the process in the public interest. The sum of the legislation and PUC codification has the following major features:

Mandatory Rate Reduction

The PUC has mandated on January 1, 2000 a retail rate reduction of 10% by the state's only utility (which we term the Incumbent Vertically Integrated Utility or IVY Power Company). Cornucopia has suffered high electrical costs, and this fact is driving industries out of the state. The citizens of Cornucopia are elated with their state government for imposing this large rate reduction on IVY Power, as its rates are highest in the nation and its power supply unreliable, with frequent shut-downs from inadequate maintenance. Three-quarters of IVY Power's board of directors have interests in entities that are IVY Power's sole-source suppliers for legal services, public relations, and the like. For the past few years, the IVY Power board compensation committee has granted the CEO annual bonuses above $5 million in spite of the utility's poor performance. It is also broadly known that the chairman of a state legislative committee has a stake in IVY Power's status quo, as he has an interest in a wholesale power marketing entity selling power to IVY Power under PURPA. It is obvious why Cornucopia's ratepayers are delighted that relief is in sight, and why the utility's management and vendors are nervous.

Unbundling of Functions

IVY Power must "unbundle" its vertically integrated functions. This means the generating assets must be separated from the transmission and distribution assets. The management of the generating assets (power plants) has to be "fire-walled" from the rest of the company -- that is, the unbundled asset must be managed as an independent entity, with separate financial statements and as a business entity distinct from the company. There can be no collusion in business planning, financing or operations -- to the extent that the power generation operation must occupy separate office facilities. The intent of this provision is to obviate any conflict of interest. IVY Power's generating assets can receive no preferred treatment in competing with other power generation sources that wish to serve IVY Power's retail market.

Auction of Generating Assets

IVY Power must place its generating assets for auction by the year 2001. The PUC will set the rules and referee the auction, including establishing the minimum value of the asset. IVY Power may bid on its own generating asset but will have no inherent advantages in the bidding. If IVY Power does win the auction, the unbundling and fire-walling provision required under the PUC's "Unbundling of Functions" will continue and be buttressed.

Uncompetitive Assets

A utility like IVY Power may have generation assets that cannot compete in an unregulated retail market. For example, it may have a nuclear plant with capital amortization costs that price its electricity above market value. Or it may have a long-term power purchase contract with a PURPA QF with high, uncompetitive rates. The difference between what IVY Power would realize from competitive rates and those it must charge for power from a nuclear plant or QF are the so-called "stranded costs" or "stranded assets."

Fairness would dictate that IVY Power's shareholders should not be penalized by the PUC's mandatory rate reduction, which would deny them a return on their investment. After all, IVY Power built its nuclear plant with the encouragement and approval of the PUC, and arguably a type of social contract exists between the state and IVY Power's shareholders. Now that the PUC has changed the rules, it seems unfair to penalize IVY Power's investors (unless IVY Power acted and managed "imprudently"). The same logic carries for QF power purchase agreements, which were imposed on IVY Power by PURPA. Thus, there is a need to deal with the inequity of the stranded asset situation.

The calculus for this involves some balancing of the requirements of mandatory rate reduction, the auction of generating assets, and the book value of the stranded asset. For some states, which introduced retail competition early, the solution was logical: "securitization." Stated simply, securitization involves the state writing a check to IVY Power for the amount of the stranded costs (or financial harm to IVY Power shareholders from the rate reduction). In the case of a nuclear plant, this could be based on the undepreciated book value of the asset, or some other number divined by the calculus of the rate reduction and auction results. The calculus of this transaction is yet to be developed, let alone understood, and the PUC has asked for more time and appropriations to make a determination.

Whatever the calculus, the check to IVY Power -- say it is $1 billion -- is backed by the full faith and credit of the State of Cornucopia, and is financed by Cornucopia-issued bonds. Bond service is accomplished by having IVY Power act as the collection agent through a "wires charge" fee to the ratepayer. Thus, the customers, who have celebrated a rate reduction and market-based rates, remain stuck with the bill for the stranded assets and will not see a real market-based rate until the bonds are paid. But securitization is a means for resolving the issue of the social contract with utility shareholders, and it moves this issue off the table.

Standard Offer (SO)

The transition from vertically integrated utilities to retail competition for power generation will take time to implement, as it moves through the steps above. Cornucopia and its PUC cannot accomplish this overnight. As a result, following the initial 12-month introductory phase (with the 10% mandatory rate reduction), a seven-year transition period begins in which the PUC mandates a transition price or "standard offer" keyed to the 10% mandatory rate reduction. Because most of IVY Power's customers are ignorant of the fine details of the transition, they need time to be educated. The probable generating entities entering the market need time to get positioned. Therefore, during the period from January 2001 to January 2008, Cornucopia has decreed that IVY Power will offer electricity at 3.5¢/kwh from 2001 through 2006, and at 4.0¢/kwh in 2007 and 2008.

The SO is important because it sets the competitive market during the transition to market-based prices. It is the SO prices that a new generating entity must equal or beat. For example, if one wishes to buy an existing nuclear plant and operate it for retail customers, the SO will determine a purchase price that will make the plant competitive during the transition period. If one is a natural-gas fired plant developer, the SO will determine the budget for pipeline development, plant capital, fuel that will keep the plant competitive during the transition.

Renewables

Cornucopia has a diminishing industry base due to rising power costs, corporate and personal income taxes, property taxes, and growing state government deficits. All of this notwithstanding, political forces in Cornucopia still favor subsidies for failed renewable energy projects, even though these projects have lead to a preponderance of stranded costs. On the other hand, nuclear power remains demonized, as does fossil generation (to a lesser extent). The politically correct energy lobby has persuaded the state legislature to continue these subsidies, and further them through wire charges and the state tax base. While touting competition and the virtues of free competition, these forces seek further succor at the ratepayers and taxpayers trough.

Let us now pause here. We have created something of a caricature with Cornucopia in order to render understandable the fundamental issues that must be addressed in a shift to deregulation. In so doing, we have oversimplified an enormously complex subject. In particular, the stranded asset discussion above is only highlighted, as are the other descriptions. But we are attempting to sort through these complicated matters with the eyes of common-sense businessmen and not government regulators and lawyers.

WHAT COMPETITION IN POWER GENERATION MEANS FOR U.S. NUCLEAR POWER

On its face, the trend toward real competition in the electrical power generation sector in the U.S. could be viewed as bad news for nuclear power. This is because estimated new plant capital costs price nuclear power well outside the competitive electricity market. But new power plant construction is what the nuclear equipment suppliers most covet because this is where they can maintain technological expertise (i.e., chauvinism) *and* lay off the costs of maintaining technological staff.

The nuclear capital cost explosion came, in large part, from custom-designing plants, and the *ad-hoc* regulatory attention that custom-designs invited from the Nuclear Regulatory Commission (NRC). Standardized designs and standardized NRC regulation, were absent in the first wave of nuclear plant commercialization (1960s and 1970s). At this moment, the entire nuclear engineering, equipment supply, fuel supply and services industries have excessive overcapacity for their served market. And at this time, the natural gas-fired power plant developers hold the upper hand in new plant construction, economically and politically.

Operating Nuclear Plants

In assessing the competitiveness of nuclear power in the new world of wholesale and retail power competition, we must look to *operating* nuclear power plants (NPPs). As we speak, utilities facing wholesale or retail competition are reviewing their nuclear plant portfolios. If an NPP is in need of major capital upgrades (such as steam generator replacements), these upgrade investments may not be justified in the face of impending competition and the utility's past performance. An adjacent decision is whether to seek a license extension from the NRC. Moreover the utility, depending upon which institutional and state legislative labyrinth it is nested, may be faced with mandatory divestments and auctions. Because the rules of the game are so geographically heterogeneous, there is no nationwide template for making such decisions. For the time being each case is treated on an *ad-hoc* basis.

From the perspective of a non-utility generator, if an operating NPP can be acquired at an acceptable price, this NPP should then be competitive. This NPP may require additional capital costs for an upgrade and license extension (if it is an older plant), but these are nominal costs compared to the expense of new NPP construction. And such an acquisition should compete handily in the competitive marketplace, even with new natural gas-fired generation. The class of plants that interest us, particularly as entrepreneurs, are those old plants that came on line in the late 1960s and early 1970s. These plants had low initial capital costs, and while they may require capital upgrades and license extensions, the economics still appear very favorable. These plants can compete at acquisition costs about equal to book value.

Some, but not all, of the plants that came on line later suffered excruciating high capital costs, occasioned by regulatory caprice, extended construction periods and high interest and escalation rates. This population cannot compete at acquisition costs equal to book value. In these cases, some form of securitization or write-off is indicated.

Partially Constructed Plants

Another class of NPPs which could be opportunities for a non-utility generator is the inventory of plants that stand partly constructed -- started but not completed because of the nuclear turndown in the 1970s. In the late 1970s, a large number of NPPs under construction were terminated because of massively escalating capital costs and regulatory uncertainties. Even though these plants are nearly complete (70-90%), the costs to bring them to operating status are large and uncertain, with the result that many owners are apparently receptive to their sale at reasonable prices.

New Plants

Nuclear power, we all know, cannot compete today with natural gas in the economics of new plant construction and power generation. There have been significant strides made with the NRC to streamline the construction and operating permit phases, and the NRC has in recent times approved standardized design licenses. Notwithstanding such streamlining, unless and until nuclear plant capital costs are significantly reduced -- by shrinking the "overnight" capital costs (through de-scoping the design) and by reducing the time-related costs (interest and escalation during construction) -- nuclear cannot compete with natural gas in the new plant market. Natural gas-fired plants can be erected in 18 months for 10-20% of the cost of a nuclear facility. Our algorithm for nuclear plant construction to compete today includes cutting overnight capital costs in half and reducing construction time to four-to-five years. Thus, for the foreseeable future, we can rule-out new nuclear designs and construction.

Our conclusions to the question, "Can nuclear power compete *now* in the framework of retail competition?" are as follows:

> *Yes,* for operating NPPs, but only if the purchase transaction significantly reduces amortization costs or the current book value is acceptably low (older plants).
>
> *Possibly,* for partially completed NPPs.
>
> *No,* for new NPP design and construction. This answer remains negative until such time when the natural-gas bubble bursts (and cheap natural gas disappears), and when we can substantially reduce nuclear capital costs through standardization and streamlined licensing.

A FINANCIAL MODEL

Management of utility monopolies have typically regarded equity (shareholder) financial return almost as an afterthought. That is because composite rates of return (debt and equity) were regulated. A utility would figure its out-year operating, fuel, and capital costs, apply a discounted cash flow method to these costs, and arrive at a levelized price (c/kwh) to cover these costs. The utility would then appeal to its PUC for permission to charge customers that price. The key point is that equity return (dividends and retained earnings) were effectively viewed as just another cost, not a financial result that should be explicitly optimized. This engendered within utility management a "cost-plus" mentality, as there was absent any compelling incentive to focus on equity return and competitive pricing as is daily faced in a normal competitive enterprise.

In the new era of deregulation, this will change. Changing the paradigm was, in fact, one of the motivations for the 1992 Energy Policy Act, which mandates that the power generation function will be competitive and no longer price-regulated. Power generators will focus explicitly on price competitiveness, cost effectiveness, and equity return -- a new set of dynamics for the industry.

The model described below reflects our perspective as entrepreneurs/ financiers contemplating the acquisition and operation of an existing NPP. It is *not* the perspective of an enterprise with a large-company mentality, with under-utilized assets which must be spread over a larger synergistic volume. It is instead a perspective that is profit-focused rather than resource-driven. The threshold question the entrepreneur asks here is: Can one make an acceptable return on investment after all operating costs, fuel costs, income taxes, outlays of equity capital and debt service, *and* with electricity prices within a competitive envelope?

We have developed a financial model for answering this and related questions. Using this model, we offer the following hypothetical case study on the purchase of a NPP.

The Rocky River Nuclear Power Plant is a 1000 Mwe unit that was constructed in ten years (the construction permit {CP} was issued in 1972, and a 30-year operating license {OL} issued in 1982). In 2012, Rocky River's original OL will expire. Since the term of an operating license originally clocked from the date of a CP award, the NRC will, through a simple letter application, extend Rocky River's OL by the term of the construction period (10 years) to 2022. A more rigorous application to extend the OL beyond 2022 could be contemplated, but for now our attention is focused on the near term. For the purposes of this financial case study, we will assume that the plant's acquisition capital is amortized over 15 years (2000-2015).

The plant was constructed for $1.25 billion. The book value using straight-line depreciation will be about $500 million in the year 2000. We will assume that the acquisition price for the plant is its book value (original capital cost less accumulated depreciation) less a "securitization" payment from the state government to the incumbent utility of $400 million. Our acquisition cost is thus $100 million.

We will assume 60% debt financing for the acquisition, with an interest rate of 8% and term of 15 years. We will further assume that each year we make routine capital expenditures of $30 million, financed with 60% debt at 8% interest. Each of these expenditures is amortized over the period from the expenditure to the year 2015. A major capital upgrade (steam generator replacement) is made in

Year Six, financed with 80% debt at 7% interest, and with a term to the year 2015. Debt service and the equity portion of each capitalization are treated as cash flows. Income taxes are treated as cash flows. Deductible costs for income tax purposes are interest on debt and depreciation (straight line), as well as annual fuel, O&M, Department of Energy (DOE) spent fuel fee, the decommissioning and decontamination (D&D) sinking fund contributions, property taxes and general/administrative costs.

Thus, the model is a purely cash flow calculation from the perspective of equity holders. The only place where depreciation enters the calculation is in the calculation of income taxes. We deliberately avoid constructing an "income" statement, which is based on the abstractions of accrual accounting. Cash is something that can be touched, while income is something that has to be defined.

With respect to the D&D fund, $500 million is estimated as the requirement at the end of plant life in 2015. The current year D&D balance (the point when the purchase is completed) is $100 million, held in escrow by the current owners. This balance will grow to $275 million by 2015 at its 7% interest rate. The shortfall of $225 million needed to fund the required $500 million is made up through annual payments to the sinking fund by the new owners.

Regarding revenues, an 85% capacity factor is assumed over the 15-year period, and a price (c/kwh) pegged at a "standard offer" of 3.5¢/kwh for the first five years and 4¢/kwh for the second five years. The final five years are priced at 4.5¢/kwh.

Table 1 shows the investment performance given the above assumptions. Note that the payback period (on the original equity investment) is about one year. The equity return accumulates to over $800 million over the 15-year period. The internal rate of return is 125%.

This clearly is a robust financial performance, and it obviously reflects the fact that the plant was acquired at a discounted value. It is not surprising that if one essentially eliminates the dominant nuclear cost component (capital cost), then nuclear can compete handily with natural gas or with the "standard offer" price allowed in the deregulating environment. On the other hand, should one substitute new plant assumptions at, say, a capital cost of $3 billion, the financial results would be dismal. However, this does not gainsay the proposition that acquiring and operating an existing NPP can (and will) be an attractive equity investment.

In the above example, if one were to use the full book value rather than the book value after the securitization offset, the results would still be acceptable. This begs the question: If, in the hands of an entrepreneur, an existing plant can be managed profitably, why do the current owners not enjoy the same financial performance? Actually, some do. But a more pervasive reason is that until now, utilities have not been compelled to deal with equity financial performance. Instead, they have been driven by an outdated cost-plus monopolistic culture.

If all of this is valid -- and we submit that it is -- the salvation of U.S. nuclear power lies with Adam Smith, not Uncle Sam. If one or more private entities acquired a dozen to fifteen plants that performed as the example above, the return-to-equity stream alone could fund the future of U.S. nuclear power, including new NPPs, fuel reprocessing, waste disposition and even a breeder program.

TABLE 1

Equity Cash Flow—Investment Performance

Year	Energy (B kwh)	Price (c/kwh)	Revenue	Fuel	O&M	Waste Fee	D&D Fund	G&A	Property Tax	Total (Annuals)	Income Tax	CapEx (Equity)	Debt Service	Annual Equity Return	Cumulative Equity Return
0			$0.00								0.00			($40.00)	($40.00)
1	7.446	3.5	$260.61	(30.00)	(80.00)	(7.45)	(8.92)	(10.00)	(8.00)	($144.36)	(44.53)	(12.00)	(9.19)	$50.53	$10.53
2	7.446	3.5	$260.61	(30.00)	(80.00)	(7.45)	(8.92)	(10.00)	(8.00)	($144.36)	(42.98)	(12.00)	(11.47)	$49.79	$60.32
3	7.446	3.5	$260.61	(30.00)	(80.00)	(7.45)	(8.92)	(10.00)	(8.00)	($144.36)	(41.39)	(12.00)	(13.86)	$49.00	$109.32
4	7.446	3.5	$260.61	(30.00)	(80.00)	(7.45)	(8.92)	(10.00)	(8.00)	($144.36)	(39.74)	(12.00)	(16.38)	$48.12	$157.44
5	7.446	3.5	$260.61	(30.00)	(80.00)	(7.45)	(8.92)	(10.00)	(8.00)	($144.36)	(38.03)	(12.00)	(19.06)	$47.15	$204.60
6	7.446	4	$297.84	(30.00)	(80.00)	(7.45)	(8.92)	(10.00)	(8.00)	($144.36)	(52.61)	(12.00)	(21.94)	$40.92	$245.52
7	7.446	4	$297.84	(30.00)	(80.00)	(7.45)	(8.92)	(10.00)	(8.00)	($144.36)	(40.35)	(12.00)	(42.49)	$58.63	$304.15
8	7.446	4	$297.84	(30.00)	(80.00)	(7.45)	(8.92)	(10.00)	(8.00)	($144.36)	(38.61)	(12.00)	(45.95)	$56.91	$361.06
9	7.446	4	$297.84	(30.00)	(80.00)	(7.45)	(8.92)	(10.00)	(8.00)	($144.36)	(36.69)	(12.00)	(49.84)	$54.94	$416.00
10	7.446	4	$297.84	(30.00)	(80.00)	(7.45)	(8.92)	(10.00)	(8.00)	($144.36)	(34.48)	(12.00)	(54.35)	$52.64	$468.65
11	7.446	4.5	$335.07	(30.00)	(80.00)	(7.45)	(8.92)	(10.00)	(8.00)	($144.36)	(48.18)	(12.00)	(59.79)	$70.74	$539.39
12	7.446	4.5	$335.07	(30.00)	(80.00)	(7.45)	(8.92)	(10.00)	(8.00)	($144.36)	(44.64)	(12.00)	(66.77)	$67.29	$606.68
13	7.446	4.5	$335.07	(30.00)	(80.00)	(7.45)	(8.92)	(10.00)	(8.00)	($144.36)	(39.21)	(12.00)	(76.87)	$62.63	$669.31
14	7.446	4.5	$335.07	(30.00)	(80.00)	(7.45)	(8.92)	(10.00)	(8.00)	($144.36)	(27.62)	(12.00)	(96.31)	$54.78	$724.09
15	7.446	4.5	$335.07	(30.00)	(80.00)	(7.45)	(8.92)	(10.00)	(8.00)	($144.36)	(80.75)	(12.00)	(7.01)	$90.95	$815.04

All currency figures are in millions of dollars.
Internal Rate of Return (equity) = 125%.

ENERGY POLICY: CAN A FREE MARKET MAKE THE "RIGHT" CHOICES?

The history of federal involvement in energy policy has not, by any reasonable measure, been exemplary. Historically, the efforts by the federal government to manipulate choices of primary energy sources -- to tilt the marketplace through R&D, direct subsidies, price supports, and similar acts of largess, have moved the energy economy farther away from Adam Smith and some sector of the industry is harmed. Witness the various federal mandates on fuel use over the years. Witness PURPA and the $40 billion of stranded costs created for electric rate payers. Witness the gas lines, dislocations and oil shortages during the federal regime of petroleum price controls.

Yet energy security is equal to economic security, which leads to national security. National security and the provision of a stable currency apparatus are two intrinsic functions of government. With a stretch, one could argue from this logic that energy is an appropriate governmental function. However, the federal government's track record in the energy sector suggests that national security is best served by letting Adam Smith have the responsibility.

A remarkable phenomenon has occurred during the past two decades -- the rise of natural gas as the dominant fuel of choice for new electric generation. Twenty-five years ago, natural gas resource base estimates (Hubbard curves) projected only seven years of forward use at the then-rates of consumption. Moreover, natural gas was regarded as a supra-convenience fuel, to be shepherded and harvested sparingly, and certainly not for mass electricity production.

The other more abundant and less versatile fuels, such as uranium and coal, were considered more efficacious for electricity production. That was the thinking then. Two things occurred since to further natural gas's role as an electricity fuel: PURPA in the late 1970's and natural gas deregulation by the Reagan Administration in the mid-1980's. PURPA gave natural gas a market advantage through cogeneration QFs, and the Reagan natural gas deregulation led to a significant increase in natural gas reserves and production. Additionally, the Energy Policy Act of 1992 gave natural gas its current power generation market opening.

The dilemma in this is that nuclear is the most efficacious fuel for generating electricity. Nuclear power remains an abundant fuel source (essentially unlimited with breeder technology) and is environmentally friendly, not only helping banish the air pollution and global warming scourge fostered by fossil fuel generation, but freeing up the more versatile and precious fossil fuels for better use. Unfortunately, a free market will not select nuclear power as long as there are cheaper alternatives, such as natural gas.

We thus believe that the nuclear industry must move forward and work smarter to compete as the playing field levels. There will be some unavoidable inequities in the transition to deregulation. Investments made by original utility investors, and later frustrated by policy and regulatory caprice, should largely be made whole. This "equitable house-cleaning" is necessary to fulfill the social contracts with original investors in nuclear units. It is also necessary before a level playing field can be established, and before a return to equity for new nuclear plants can be generated to fund the future of nuclear power. Put bluntly, there is a class of plants that came on line in the 1970s that suffered from government-triggered high interest rates, construction delays and escalation rates that preclude them from competition today. If the difference between these plants' capital costs and market value can be reconciled through a one-time securitization program, then these plants can be recovered and operated successfully. This inures to the public's benefit and to the future of nuclear power.

Should we continue toward Adam Smith or reverse to the command-and-control energy policies which have wreaked such havoc in the past? We are not persuaded that Uncle Sam will get it right this time either. The performance of DOE over the past 10 years in executing its fundamental statutory responsibilities in nuclear waste management and uranium enrichment confirms this. On the other hand, can we expect a free market to get it right if Adam sets the regulations rather than Uncle? We think they will.

SUMMARY AND CONCLUSIONS

The current deregulation and restructuring of the power generation component in the U.S. electric utility industry offers unique challenges and opportunities for the nuclear industry. The opportunities should be seized and not only used to stop the decline of the nuclear industry, but as the platform for a renaissance of nuclear power. This will require some fundamental changes in industry attitudes and aptitudes -- particularly a shift from heralding technological development to becoming more conscious of, and competent in, the financial performance of the business. Adam Smith, not Uncle Sam, must be recognized as the savior of the future for nuclear power.

This renaissance cannot be approached from the old cost-plus, big company mentality and culture. Over the next few years, a number of nuclear plants will be bought and sold, and companies that seek to lay-off their large fixed costs on a newly-acquired NPP will fail. Should this occur, nothing fundamental will have changed. Vertically integrated monopolies will simply be replaced by horizontal monopolies.

An entrepreneurial approach, stressing economic and financial fundamentals, driven top-down, can succeed. Some plants, if acquired at book value, can readily compete. Other plants, especially newer facilities, will be unable to compete without securitization or writeoffs. Securitization is not Adam Smith's invention, but it can lead to a competitive market as a one-time transitional event necessary to level the playing field and fulfill outstanding social contracts between utility shareholders and the state governments. This is a much smaller price to pay to reach the free market promised land than to stay in the dark valley of a regulated world.

Transition periods are never easy, and the mental and behavioral adjustments required of those accustomed to the old utility culture, and who now must operate in a new world of competition, will be tough for many. Look only at the executive displacements and career tombstones wrought by the deregulation of the airline, natural gas and telecommunications industries as they made this transition. This is a time for fresh thinking, entrepreneurship, risk-taking and vision in the nuclear industry. Those who have this mental outlook and board the deregulation train now leaving the station -- and who leave their old regulatory and cost-plus intellectual baggage behind -- will be in for an exciting, profitable and satisfying ride. All aboard?

THE FRENCH GERMAN PROJECT OF THE EUROPEAN PRESSURIZED WATER REACTOR (EPR)

François Bouteille, Executive Vice-President

Nuclear Power International, NPI
6 cours Michelet, 92064 Paris la Défense -France

INTRODUCTION

The basic design of the European Pressurized Water Reactor (EPR), which began in February 1995, was completed in June 1997. Electricité de France, the participating German utilities, Framatome, Siemens, and Nuclear Power International (NPI), a joint subsidiary of Framatome and Siemens, then decided to engage an additional design phase: the basic design optimization phase. Started in June 1997, this phase will end in December 1998. The status of the project is summarized hereunder.

The year 1998 was also marked by a deeper political reflection for integration of a lead EPR nuclear power unit into the context of long-term French energy policy. The idea that it is necessary to decide the construction of this lead unit as early as possible has made progress, particularly in influential political circles. This decision will take on all its significance if it is placed in a framework of cooperation between French and German industries, in line with a possible enlargement to other European countries interested by this program. EPR development has therefore reached a crucial stage, and the orientations that will be set soon will enable giving a new impetus to this next-generation reactor program.

EPR DESIGN PHILOSOPHY

Safety Approach

NPI's overall approach was defined to meet the requirements set up by GPR and RSK (Advisory Groups which are giving advice respectively to the French and German Safety Authorities). Their common document is entitled "GPR-RSK Proposal for a Common Safety Approach for Future Pressurized Water Reactors". It was issued in 1993.

A two-fold safety strategy has been pursued to further enhance the already very high level of safety of the existing Framatome and Siemens technologies. The objectives are, first, to achieve improved accident prevention. Second, simultaneously with a drastic reduction of the probability of severe accident scenarios (up to core melt), additional features, mainly concerning containment, are being implemented to mitigate the consequences of such accidents.

Preparing the Ground for Renewal of Nuclear Power, edited by
Kursunoglu *et al.* Kluwer Academic / Plenum Publishers, New York 1999.

The overall aim is to ensure that even an accident of extremely low probability would not require any stringent countermeasures, such as evacuation or relocation of the population beyond the immediate vicinity of the plant. This objective complies with the amendment to the Atomic Law enacted in Germany.

Probabilistic risk assessment is an essential addition to the deterministic basis of the EPR safety approach. It is used to demonstrate a well balanced contribution of the relevant accident scenarios and transients to the overall probability of severe accident scenarios, and to demonstrate a sufficiently low probability for accident that could lead to large radioactive releases.

The design objective is to meet two major safety goals with adequate margins: the integral core melt frequency, considering all plant states and all types of events (internal events, internal and external hazards), is smaller than 10^{-5}/reactor-year (set forth in INSAG[1]-3) and the probability of large radioactive releases is smaller than 10^{-6}/reactor-year.

Performance

Taking into account the feedback of experience from the utilities participating in the project, ambitious objectives were defined to minimise as much as possible the cost of operation and to meet the global requirements of the operators. Among the most significant objectives one can mention:

- The search for the lowest generating cost that has led the Project Directorate to decide a power uprate of the plant. The targeted value is 1750 MWe.
- A large flexibility of fuel management to encompass long fuel cycles with out-in or in-out core loading patterns and the capability to include 50 % MOX recycling.
- A very high availability (more than 90 %) which is possible thanks to the capability to perform fuel reload within very short outage, possibly in 19 days. A strong emphasis is being put in the design to facilitate the maintenance of the plant. In this respect, the four-train architecture allows to perform preventive maintenance on some systems during power operation. This is an important contributor to the capability to ensure short outages. Furthermore, the design will incorporate features, which are devoted to limit the reactor trip occurrences, such as the partial trip.
- A flexibility to allow permanent adaptation of the electrical output of the plant to the demand of the grid. The ability of the plant to perform load follow is directly inherited from the French and German PWRs which were designed and are performing daily load follow and continuous frequency control.

Competitiveness

The cost target of the EPR is defined in the European Utility Requirements. They call for a generating cost advantage of 15% over current plants for base load operation. The preliminary evaluation shows that the generating cost of the EPR should be around the lowest value which is expected to be achieved by the combined cycle plants even taking into account their potential for greater efficiency. Even if it is true that the price of fossil fuel is very low and that the trend is not in favour of nuclear energy, the policy makers should take into

[1] INSAG : International Nuclear Safety Advisory Group

account the large potential of price instability of gas and oil due to political reasons. It is not necessary to look far away in the history to get this reminder. Furthermore, the orientations decided during the last Kyoto conference have the clear consequence that, on a general basis, existing Nuclear Power Plants can only be replaced by new nuclear plants.

The highest possible size of the unit has been selected to minimise the operation and maintenance cost. From known experience, it is well known that these costs are rather independent from the unit size and can be regarded as fixed cost. The technical requirements for the core design are specified in a way that the plant is able to generate the rated power as specified above with the lowest cycle fuel cost. This design provides the utilities with a wide range of fuel cycling flexibility and opens up the choice for an optimised back end strategy in terms of disposal costs. In addition all cores shall be designed so that refuelling schemes for cycles between 12 and 24 months are feasible. This provision contributes also to the fuel cycle flexibility and provides margins for optimisation. Furthermore, the possibility for long cycles reduces the expenses for maintenance works and increases the availability of the plant and consequently its competitiveness. The burnup target for the fuel assemblies has been set to 60 000 MWd/t. This burnup target will provide a high usage factor of the fuel and thus contribute to favourable fuel cycle costs.

The 1750 MW electrical output of EPR is produced out of a thermal core power of 4900 MW. This capacity is produced by a core consisting of 241 fuel assemblies of the well proven 17x17-25 design but with an active length of 420 cm, so that the average linear heat rate remains around the well proven value of 178 W/cm. Special measures on the secondary side of the steam generators provide the EPR with an increased efficiency of more than 36%. This efficiency is gained by providing, as for the N4 French units, the steam generators with an economiser which results in an increased main steam pressure and by optimising the cold end of the plant for which the conventional island supplier takes the responsibility. As explained previously the improvement of the availability is one of the key objectives of the project and contributes directly to lower the cost of production.

MAIN DESIGN FEATURES

Essential design and operating data

The main reactor core and reactor coolant system operating data are listed in Table 1. The main reactor coolant system components were enlarged relatively to the existing practice: a larger reactor pressure vessel will accommodate the large core size. The pressuriser and the steam generator secondary sides are enlarged to improve the plant transient response. In case such as the total loss of feedwater flow, the grace period available for appropriate corrective actions by the operators will be increased.

Primary and Secondary side safety systems

The development has been based on the following principles.

Simple and straightforward design: Important safety functions, to mitigate abnormal occurrences are assured by separate systems in a straightforward operating mode. Combination of several safety functions in a complex system design is avoided. Thus

Core thermal power	4900	MW
Rated electrical power (gross)	≈ 1750	MW
Number of loops	4	
Operating pressure	155	bar abs
RPV inlet / outlet temperature	289.8 / 330	°C
Total vessel mass flow rate	23148	kg/s
Main steam pressure	74.6	bar abs
Number of fuel assemblies	241	
Number of rod control cluster assemblies	89	
Fuel assembly array	17 x 17 - 25	
Active height	420	cm
Average linear heat rate	178	W/cm
Enrichment (max) U_{235}	4.9	%
Batch discharge burn-up	60	Mwd/kg

Table 1 - Design and Operating Data

operating and maintenance staff will get a clearer understanding of plant states in all conditions.

Separation: Different redundant trains of all safety systems are installed in four separate layout divisions for which a strict separation is ensured so that common mode failure for example due to internal hazards can be ruled out.

Functional diversity: A reduction in common mode failure potential is obtained by design rules ensuring the systematic application of functional diversity. In this approach, complete loss of all redundant trains of a safety system is postulated and other safety systems with functional diversity are relied upon to maintain the plant in safe conditions.

Redundancy: A four-train redundancy is used for the major safety systems ensuring emergency core cooling, steam generator feedwater supply, and associated support functions. Some other systems are arranged in a two-train configuration (e.g. spent fuel cooling pool, Extra Borating System : EBS). The four-train concept, associated with a four-loop Reactor Coolant System, contributes to a simple and straightforward system design as mentioned above. It provides flexibility in adapting the design to maintenance requirements, thus contributing to reduce the length of the outages. The residual heat removal at cold shutdown is ensured by the Low Pressure Safety Injection System. The location of the main active components outside of the Reactor building simplifies the maintenance operations.

Mitigation of Severe Accidents

As already mentioned, additional features are implemented to mitigate the consequences of low pressure core melt accidents. This is the consequence of the basic safety rules adopted by the French and German Safety Authorities. Basically they can be summarised by the three following objectives:

- Achieving a significantly lower core melt probability by appropriate prevention means,

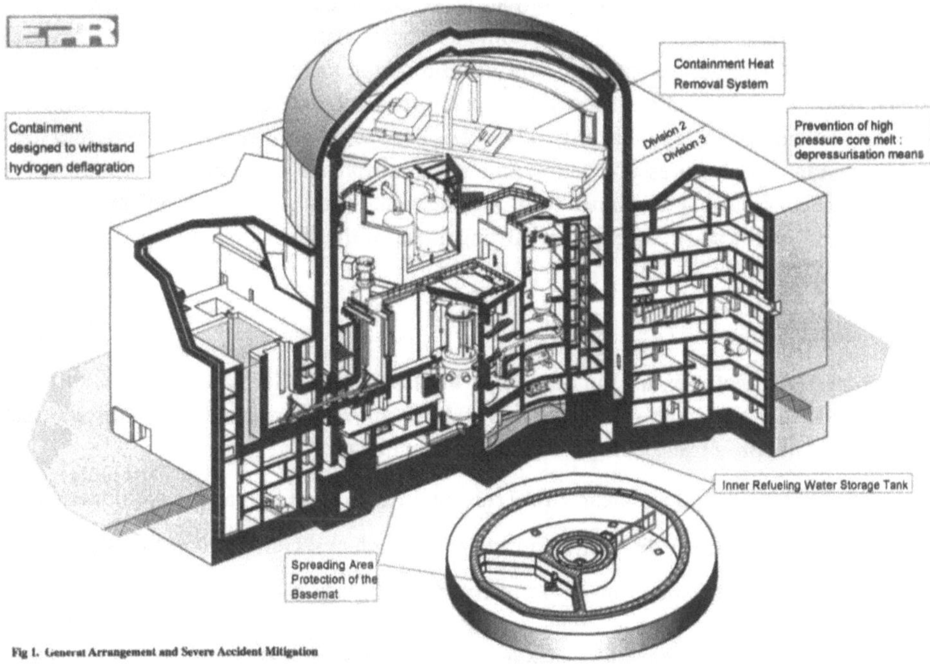

Fig 1. General Arrangement and Severe Accident Mitigation

- Achieving the "preclusion" of accidents liable to cause early containment failure such as core melt under high pressure conditions,
- Achieving a major reduction in the radioactive releases, which could result from low pressure core melt accidents.

As a consequence of this safety requirements, one specific feature of the EPR is the lower containment part which is designed to stabilise and to retain the molten core (see Figure 1 for illustration). The reactor pressure vessel may fail from erosion by the mixture of molten metal and oxide debris. Steam explosion is prevented by minimising the amount of water where the corium is spread. The prevention of the molten core-concrete interaction is obtained by spreading the corium in a dedicated spreading chamber. This original EPR feature consists in a large area outside the reactor pit.

The reactor pit and the spreading chamber are connected via a discharge channel, which is closed to retain the melt for a certain time in order to accumulate the melt in the pit.

The spreading area is connected with the In-Containment Refuelling Water Storage Tank (IRWST) with pipes for water flooding after spreading. These pipes are closed during normal operation and accident conditions by plugs, which will be molten by the corium after spreading.

A containment heat removal system is dedicated to limit the containment pressure in the specific case of core melt ; it consists of a spray system which decreases the containment pressure nearly down to atmospheric pressure. Even if all other systems are not yet available, the containment design pressure grants a grace period of at least 12 hours after the accident before the operator may need to actuate this spray system.

The reactor containment, the third and ultimate barrier under the defence-in-depth concept, is particularly important in mitigating the consequences of severe accidents.

A cylindrical, double-wall containment design has been adopted for the EPR. A prestressed concrete cylindrical wall and a reinforced concrete basemat constitute the inner containment. The outer wall is formed by a reinforced concrete cylindrical wall, which serves as protection against external hazards. The space between the inner containment and the outer wall is maintained at a pressure below ambient atmospheric pressure, in order to collect all possible air leaks through the inner wall or penetrations and filter them before release to the environment, via the stack. The general arrangement of the plant is shown on Figure 1.

CONCLUSION

Cooperation between all the players in the French and German nuclear entities is now well-established: safety authorities, utilities and industry. It broadens the market accessible to the EPR, and important results have already been achieved:

- Harmonization of the nuclear safety rules in France and Germany is well under way.
- The standardization effect will benefit both French and German industries.

The new political situation in Germany is modifying significantly the overall picture. But the first reactions of the main actors indicate that this will not stop the fruitful cooperation between France and Germany. Even if the new government in Germany has repeatedly stated that they will go for "a new, sustainable energy mix without nuclear power", it remains to be demonstrated that this miracle solution actually exists. The difficulty is of course to define a solution which can meet the commitment defined in the Kyoto conference in terms of reduction of CO_2 production.

We have to look ahead confidently to 1999, which will be a critical year for the launching of the first EPR nuclear power plant.

ASSURING THE COMPETITIVENESS OF NEW NUCLEAR PLANTS IN A DEREGULATED U.S. MARKET

George A. Davis

ABB Combustion Engineering Nuclear Power
Windsor, CT, USA

SUMMARY

Deregulation of the U.S. electric power industry will dramatically impact the way in which future plant buyers evaluate technologies available for adding new generating capacity. Compared to the regulated utilities of the past, future plant buyers will tend to be more averse to investing in capital-intensive technologies. In addition, new cost projections for generating electricity from technologies based on fuel sources other than nuclear energy are predicting substantial cost reductions over the next twenty years.

As recently as 1995, industry studies had concluded that new nuclear energy plants would be competitive with coal and natural gas based alternatives in the U.S. market, if a nuclear plant's life cycle cost could be held below 4.3 cents per kilowatt-hour. Thus, the central economic goal of the Advanced Light Water Reactor (ALWR) program was set at this value. (The ALWR program was a joint government/industry effort to develop and license a new generation of standardized reactors that combined all of the lessons learned from the first generation of nuclear plants, as well as resolution of regulatory/safety issues that arose after the Three Mile Island accident.) However, new studies are now showing that, to be competitive in the long term (20 years from now), new nuclear energy plants in the U.S. may need to produce electricity at total costs below 3.0 cents per kilowatt- hour -- assuming that the government does not step in and impose a carbon tax (or other disincentive) for the use of fossil fuels. Since operating and fuel costs for nuclear plants are proving to be quite competitive with other fuel sources, the major impediment to long term competitiveness of new nuclear plants in the U.S. is the capital cost component -- which may need to be reduced on the order of 35%. Such a substantial reduction in capital cost would require a fundamental reevaluation of the industry standards and regulatory bases under which nuclear plants are designed and licensed.

As is discussed below, such a reduction in capital costs can likely be achieved by combining three major process tools: (1) application of advanced technologies developed in other industries (particularly computer technology) through all phases of design, licensing, fabrication, construction, and operation; (2) use of probabilistic risk assessment as a design tool to simplify designs, and (3) extensive application of risk-based regulation as a means to streamline the regulatory requirements and process. It would likely take more than a decade to fully develop and implement these processes -- resulting in a new generation of pre-licensed nuclear plant designs that would be economically competitive in the long term, deregulated U.S. power market.

The Impact of Deregulation on the Power Industry

With deregulation of the U.S. power industry, the decision on which type of new plant to purchase (nuclear, coal, natural gas, hydro, wind, solar, etc.) will be made by an unregulated independent power producer. The only question to be answered will be: "Which technology is most likely to make the most profit, with the least risk?" Thus, buyers will be seeking the lowest cost, with the least financial uncertainties. Every factor and unknown will be converted to dollars and risks on those dollars. The social issues that were previously factored into such decisions and considered between regulated utilities and their public utility commissions (e.g., energy independence, aesthetics, impacts on the local economy, etc.), will no longer be considered -- except to the extent that they are covered by minimum government standards, incentives, or penalties (e.g. EPA regulations, NRC regulations, tax incentives for using certain fuels, tax penalties for using certain fuels, etc.).

In the regulated marketplace, utilities based their decisions on total life cycle costs (capital, operating, and fuel costs combined). If the assumptions in their original analyses later changed (e.g., new NRC requirements were imposed), then their public utility commissions would usually allow them to add the increased costs into their prices, as long as their decisions had been considered prudent. In general, the regulated utilities were exposed to less risk on their capital investments -- in return for accepting limited profit margins.

In the future deregulated U.S. market, however, there will no longer be a direct linkage between the cost to produce electricity and the price that can be charged for it. Price will be determined by the competitive marketplace, without the opportunity to increase prices to cover unanticipated costs. As a result, buyers of power plants in a deregulated marketplace will be much more concerned about investing substantial amounts of up-front capital in a project -- with the risk that subsequent events could unexpectedly make the plant become noncompetitive before the capital is repaid.

For example, if a power producer chooses a natural gas plant, the capital costs can be repaid in about ten years from the time that the decision is made. A nuclear plant (of the same generating capacity) would require four times the capital investment and well over twenty years to repay. Thus, to consider a capital-intensive plant (e.g., nuclear), the power producer will likely demand a higher return on investment, to justify the longer term risks -- much like 30 year bonds must provide higher returns than 10 year bonds, to attract investors. In a market where natural gas prices are considered stable or declining, then the premium required for nuclear investments could need to be substantial. On the other hand, if gas prices are expected to rise significantly or are erratic, then the premium required for nuclear might be small.

Another uncertainty factor resulting from the longer repayment period is the increased risk that new technologies may become economically competitive and enter the marketplace and further lower electricity prices that can be charged, before the capital investment on the nuclear plant can be repaid. (For example, in 1978, almost no one would have anticipated that natural gas would become the most preferred power generation option in 1998.) In the end, future investors may simply require that the capital costs must be recovered in a much shorter time period than previously allowed in the regulated marketplace -- thus, creating even more pressure to hold down capital costs on future nuclear units.

The issues that are unique to nuclear (e.g., spent fuel disposition, NRC actions, public perceptions of safety, decommissioning standards, etc.) will all be converted to financial costs and financial uncertainties. They will then be factored into the economic decision process for selecting which technology to use. Even if the best-estimate financial comparisons show the nuclear energy option to be competitive, the financial uncertainties for the nuclear issues could still be large enough to tip the scales against the nuclear option. (This has, indeed, been the case in some of the recent decisions to shut down operating plants in the U.S., before their NRC licenses have expired.) Therefore, reducing financial uncertainties will be just as important as reducing the nominal costs.

The Current Status of Nuclear Energy's Economic Competitiveness in the U.S.

For new nuclear plants in the U.S., the ALWR designs are the most advanced products available today and, thus, provide the best reference for considering the industry's current state of the art. When the ALWR program was initiated in the mid-1980s, economic goals were among the first to be established. The most important economic criterion was the total life cycle cost, based upon the expectation that the ALWRs would be used by regulated utilities. In 1995, the economic goals for the ALWRs were updated. Estimates, at the time, indicated that new nuclear energy plants would be competitive with coal and natural gas based alternatives in the U.S. market, if a nuclear plant's life cycle cost could be held below 4.3 cents per kilowatt-hour. Thus, the central economic goal of the ALWR program was set at this value.

However, new studies are now projecting substantial reductions in the costs for generating electricity from fossil fuel sources, over the next twenty years. To be competitive in the long term U.S. market, nuclear energy plants may need to produce electricity at costs below 3.0 cents per kilowatt- hour -- assuming that the U.S. government does not step in and impose restrictions or taxes on the use of fossil fuels. (Although it is difficult to estimate anything twenty years in the future, including electricity generation costs, the trend toward lower plant costs and higher plant efficiencies is very clear.) In addition, as explained earlier, deregulation of the power industry will result in greater emphasis upon minimizing capital costs. As a result, nuclear energy is likely to become noncompetitive for future plants in the deregulated U.S. market -- unless there is a concerted effort to reduce capital costs for nuclear plants to match the reductions being achieved for non-nuclear technologies.

Although the ALWR program was just finished in 1997 for the two large, evolutionary designs (ABB-CE's System 80+ ALWR™ and GE's ABWR), efforts during the last several years were devoted to gaining NRC approval and certification of the designs. The major design features were developed during the latter 1980s and early 1990s-- in parallel with the industry's development of the ALWR Utility Requirements Document. In the regulatory environment that existed at that time, there were very few opportunities to actually reduce the regulatory requirements for the ALWRs. Except for a handful of "optimization issues", the ALWR designs were developed by taking existing NRC requirements and adding to them additional requirements to address the NRC's list of unresolved safety issues, as well as severe accident issues that emerged after the Three Mile Island accident. Probabilistic risk analyses were used to identify weaknesses in the designs, which were then corrected. On top of this, there were a number of ALWR requirements for design features that would make the plants easier to operate and maintain.

While the industry was, on one hand, trying to develop design features to reduce capital costs, there were substantial pressures to implement new requirements that added to capital costs. The philosophy was that higher overnight capital costs would be offset by shorter licensing/construction schedules and reduced (and more certain) operating & maintenance costs. This was acceptable because the primary economic criterion was based upon total life-cycle costs -- the economic criterion of regulated utilities.

However, with the power generation industry moving to a deregulated marketplace and with the cost estimates for alternate technologies declining, there is now a clear need to substantially reduce nuclear plant capital costs. Based on new industry comparisons, the capital costs for future nuclear plants in the U.S. market may need to be on the order of 35% below the costs currently estimated for the ALWRs. This would place nuclear plant capital costs in the same general range as capital costs for coal plants. Development of new designs to produce cost reductions of this magnitude will require the use of new, innovative design tools that were not available during the 1980's. Very importantly, to implement the design changes, there will need to be a major shift in the criteria and procedures by which new nuclear plant designs are licensed by the NRC.

Applying the Computer Revolution to the Nuclear Industry

During just the last decade, the revolution in computer technology applications has dramatically changed the way in which American industry does business -- including all aspects of communications, planning, management, engineering, testing, manufacturing, construction, operation, etc. Unfortunately, compared to many other industries, applications of computer technology have not been as fully utilized by the nuclear industry -- because of a lack of new plant orders in the U.S. and because of the difficult regulatory hurdles that must be overcome to implement this rapidly advancing technology.

The automotive industry is an excellent example of what can be achieved. Computer technology allows the designers to produce cars with unibody construction that is stiffer, while using less steel and while having crumple zones built in, to protect passengers during an accident. It allows the implementation of anti-lock braking systems and traction control systems to help prevent accidents. Computer technology (e.g., robotics) reduces fabrication costs and allows engineers to pre-plan changeovers of assembly lines from one model to the next. It allows the inventory of parts to be managed so that assembly lines are not delayed while waiting for shipments. Computer technology warns the automobile's owner when there are problems and when maintenance is required. It then helps the maintenance technicians to quickly identify the source of the problem and to repair it.

Compared to the automotive industry, the nuclear energy industry has only scratched the surface of applying computer technology as a tool to simplify plant systems and structures. For example, the technology used for much of the basic equipment included in a nuclear plant (e.g., pumps and valves) has not changed significantly during the last twenty or more years. If plant designers needed to increase the reliability of the system in which that equipment was used, then they had to build in redundancy -- by adding additional trains of identical systems, as backup. Sometimes, three or four trains of identical systems were included, just to assure the reliability of a single safety function. When concerns were raised about common mode failures that might disable all of the identical systems at the same time, then diverse backup systems, using different designs, were added. On top of all of this, there was a need to employ extremely high quality assurance standards that were unique to the nuclear industry. All of this redundancy, diversity, and quality assurance has dramatically driven up the costs of nuclear energy.

Today, however, computer technology can now be developed to make equipment in future nuclear plants smarter (e.g., pumps and valves with self-monitoring, self-diagnostic features built in). Smart equipment would be more reliable and, thus, improve the overall reliability and safety of the system in which it is used. The impact on nuclear plant design could be dramatic. By addressing the question of reliability at the root-cause level, it should be possible to reevaluate the need for so much redundancy, diversity, and quality assurance. (In the case of quality assurance, it may be possible to adopt standards from other industries, which have improved substantially in recent years.) The resulting cost savings -- both capital and operating costs -- could be dramatic. Best of all, the savings could be accomplished without sacrificing safety.

Besides inclusion in the plant design itself, computer technology can be applied to all stages of the design, fabrication, construction, operation, and maintenance processes. For example, a computer-based Information Management System (IMS) with open architecture could be developed and used, industry-wide, by designers, plant owners, equipment suppliers, regulators, etc. Such a standardized IMS would allow plant designers in different companies (even in different parts of the world) to work together on a specific plant design -- instantly sharing information as it is being developed. The costs and schedule for managing, sharing, and providing quality-assurance of design data would be dramatically reduced. Very importantly, the standardized IMS would allow plant buyers and operators to later have complete access to the design information on the plant. Even NRC regulators could have access to the databases. Such a computer-based system would also avoid the need for design reconstitution programs, years after plants have already been constructed (which was recently a major and costly effort for the fleet of operating nuclear plants in the U.S.).

Computer technology can also be applied to the construction process. Besides actions to reduce the actual cost of each construction activity, anything that shortens the overall construction schedule will provide substantial savings in the interest charges that accumulate during the construction period. A three dimensional model of the plant could be constructed on computer -- with problems being identified and corrected, before any actual construction begins. If available during the design phase, such a model could even provide feedback, to identify design changes that would further reduce the construction schedule and costs. Critical path scheduling for construction activities could be developed and maintained on computer -- even allowing plant constructors to quickly recover from any unanticipated delays. Material and equipment ordering could be improved by computer applications. Equipment costs could be reduced by applications of computer technology to the fabrication processes -- perhaps, even to the point that some equipment currently manufactured overseas could be competitively manufactured in the U.S. again.

These are just examples. There are thousands of activities performed during the design, licensing, fabrication, construction, operation, and maintenance of a nuclear energy plant. The cost of every one of them could be reduced through the application of computer technologies that did not even exist ten years ago.

A More Economic, Risk-Based Design and Licensing Process

Ultimately, what will be needed is a new generation of nuclear plant designs that are substantially lower in capital costs. This can be accomplished by first developing a design process that goes through an entire plant's design and determines how each system or structure can be simplified, while still providing levels of safety and reliability that are comparable to, or better than, what was achieved for the ALWR designs. To simplify the designs, however, the process must reevaluate the current hodgepodge of nuclear plant design & licensing criteria that have been produced over the past forty years. In essence, the new process should use Probabilistic Risk Assessment (PRA) as a design tool to determine how each system or structure can be simplified, while maintaining a reliability level (which also means maintaining a safety level) that is at least as good as the original design. The reevaluation process should identify and preserve the design & licensing requirements that truly contribute to reliability and safety. More importantly, the process should identify the design & licensing requirements that unnecessarily complicate the design, without significantly contributing to the system's or structure's overall reliability. Such a process would be most effective if combined with the introduction of smart equipment, which will increase reliability at the component level.

As an example, let's look at the design of the Safety Injection System (SIS). Beginning over thirty years ago, a great deal of deterministic regulatory criteria has been developed for the SIS, based upon a postulated event that is now known to have less than a one-in-a-billion chance of occurrence: an instantaneous double-ended guillotine pipe break, in the worst location, with the worst single failure, with the worst initial conditions, with the worst operator response, with the worst coolant-radioactivity conditions, with the worst containment leakage, etc., etc. Industry standards and NRC regulatory requirements for the SIS evolved in a patchwork of documents that were generated or revised every time someone thought of a new concern, there was a new problem at an operating plant, or something was found during maintenance.

However, after the first full blown PRAs were performed in the 1970s (e.g., WASH-1400), it was recognized that the most catastrophic events imaginable were not the events most likely to threaten public safety. The double-ended guillotine pipe break was found to be of such low probability that, by the early 1980s, the NRC's Materials Branch acknowledged that ductile pipe would "leak before break" and could not pose a real threat -- as long as there was a leakage detection system. On this basis, the NRC allowed "leak before break" to be credited, in satisfying some of the *new* requirements that NRC was then imposing; however, the double-ended guillotine pipe break was still maintained as the basis for the already established NRC requirements -- which had served as the design basis for virtually everything in the plant (e.g., the safety injection system, containment, etc.). This obvious inconsistency in regulatory requirements was justified by NRC, and accepted by industry, as providing an added safety cushion, to cover the unknown.

In a young industry -- lacking a wealth of operating experience and data -- an added safety cushion, to cover the unknown, was not unreasonable. Furthermore, in a regulated power marketplace, the added requirements could be tolerated because plant owners could usually pass along the costs of satisfying the NRC requirements to ratepayers. However, in a deregulated power market, continuing the use of design features that don't truly add to safety and reliability are resulting in nuclear plant designs that will not be cost competitive in the long term, deregulated U.S. market -- and, therefore, will simply not be purchased.

The state of the art for PRAs (including the database of operating experience) is now sufficiently mature that we should be able to hone in on the industry standards and regulatory requirements that truly add to reliability and safety. Although the final decisions must always be made by humans, the decision process should now be able to rely more heavily on risk-based inputs. The nuclear energy industry and the NRC are already working together to apply risk-informed regulation to the licensing of existing plants. Though still in the early stages, this industry/NRC effort is making a great deal of progress and promises to offer significant benefits for the existing plants. However, these efforts are focused upon licensing requirements that affect the operation and maintenance of existing nuclear plants. What is needed, beyond the current effort, is to apply a more aggressive risk-based approach to industry standards and regulatory requirements, as well as the regulatory process -- focusing upon those that affect the design and licensing of new plants, rather than just the operation and maintenance of existing ones. The new plant effort should complement the existing industry/NRC effort -- not compete against it.

If we applied a more risk-based regulatory approach to the Safety Injection System example mentioned above, then it is likely that we could greatly simplify the design of the system. For example, we could specify smarter pumps and valves that provide more diagnostics and warnings to the operators -- increasing the reliability of each train. Combined with more realistic design bases, we could likely develop a two-train SIS that is just as reliable as the four-train design currently found in the ALWR designs -- with each train being simpler than those in the current designs. Even if each individual train turned out to be slightly more expensive, elimination of two trains would still dramatically reduce the overall cost of the SIS. More importantly, elimination of two trains would dramatically reduce the costs of the structures required to house the SIS, as well as the other systems that support it. It is also likely that we could then simplify the system's operating, maintenance, and testing procedures. A similar approach could be applied to containment design, etc., etc., throughout the entire plant design. Very importantly, this could all be done without sacrificing safety.

The Need for a Long Range Research & Development Program

It would likely take more than a decade to (1) develop the process tools described in this paper, (2) implement the changes to the industry and regulatory infrastructure that would be needed, and (3) develop and license a new generation of nuclear plant designs that utilize the new tools. To be manageable, it should be planned in phases -- with decisions on whether and how to proceed to the next phase based on the experience gained in the prior phase. The initial phase could even be broken into smaller activities. Much of the initial work could be done generically, independent of any particular reactor design (or even reactor type) -- as long as there is an overall master plan (i.e., a technology roadmap), to assure that all pieces of the puzzle are being addressed and coordinated.

Such a major program need not -- and should not -- be taken on by U.S. government and industry alone. International collaboration should be encouraged. For example, the Korean nuclear industry is currently developing it Next Generation Reactor -- the KNGR. The KNGR design will reflect the collaborative efforts of Korea's participation in the U.S. ALWR program, as well as its own efforts to introduce advanced technology and to reduce capital costs. The burgeoning Chinese nuclear industry also has strong incentives to collaborate in a U.S. program that aims to introduce more advanced technologies and to reduce costs. With active nuclear construction programs, the Asian nations can also provide near term opportunities to implement some of the technological advancements developed in the U.S. program -- which would also provide valuable feedback to the U.S. program and could result in valuable mid-course corrections along the way. Last but not least, the European nuclear industry is facing the same economic competitiveness pressures as are being experienced in the U.S. Collaboration with the U.S. program should be beneficial to both.

The goal should be to have a new generation of nuclear plant designs licensed and available for the marketplace by the time that (1) the dust has settled from deregulation of the power industry and (2) the new independent power producers have become firmly enough established that they can begin consideration of substantial additions to their baseload generating capacity -- by utilizing longer range capital investments. The year 2010 is probably a reasonable and achievable date to assume for the U.S. market. As the marketplace develops, assumptions and schedules will surely change and adjustments in the proposed program will need to be made, accordingly.

The program should also seek to standardize the plant designs, the equipment used, the processes employed, and the design/regulatory requirements, to the extent possible. Standardization is essential to keeping nuclear plants competitive, by allowing fixed costs to be spread over many units. Ideally, a company (or group of companies) should be able to market a single standardized design in any country around the world. In an increasingly global marketplace, international standardization must become the industry norm. Individual custom designs will not be economically competitive.

Enabling Conditions That Are Required

Several enabling conditions would be required to carry out such a program:

1. Government and industry will need to work together to convince the public that it is possible to substantially simplify the design and licensing requirements for new nuclear plant designs, by introducing smarter, more advanced technologies -- without sacrificing safety. The first step will be to convince the various government and industry organizations, themselves. When the ALWR program was initiated in the mid-1980's, it would have been impossible to suggest such a radical shift toward risk-based criteria for the design and regulation of nuclear plants. However, because of issues being addressed for operating plants, the industry and NRC are now beginning to make significant progress in developing a risk-informed, performance based regulatory approach. Looking beyond currently operating plants, industry and government should start to push the risk-based envelope to its logical conclusion for developing the licensing requirements and procedures for new plants. Many in Washington are becoming sensitive to the fact that there will eventually be no nuclear industry in the U.S., if we don't start to reform the regulatory requirements and procedures.

2. The R&D effort to carry out such a program will require the collaborative efforts of NRC, DOE, and all aspects of the nuclear industry (utilities, EPRI, A/Es,

vendors, and equipment suppliers) -- as well as participation by universities and the national labs. Such a major program will require the input of a lot of creative thinkers who are not saturated with a culture that is based on the existing requirements. On the other hand, it will also require the experience of those that know why all of the existing requirements were put there in the first place. NRC should be involved from the beginning -- recognizing, however, that a degree of regulatory independence must be maintained. Besides providing creative thinkers, participation by universities will also help to maintain the educational infrastructure that will be needed to support a long term nuclear energy option. This would further help to entice students to enroll in the nuclear energy field. Participation by the national labs would also play an essential role in providing creative thinkers, as well as providing access to many of the advanced computer technologies that are now being applied in defense and other industries. The labs can also provide the facilities needed to carry out much of the research & development needed for this program -- including the performance of test programs. Furthermore, downsizing has reduced the nuclear industry's resources available for R&D programs. The universities and national labs can help to fill this gap.

3. Full implementation of the program described above could take more than a decade to complete. Substantial resources will be required of both government and industry. Long term commitments to the program will be essential. To make the most of limited resources, there must be an atmosphere of openness and cooperation -- organizations must be willing to share information and accept new ways of doing business. Embedded industry and regulatory cultures will need to change. Standardization must be maximized.

4. Government and industry must continue to make progress in resolving issues that affect the currently operating nuclear plants: risk informed, performance based regulation; streamlining of regulatory reviews; spent fuel disposal; decommissioning requirements; plant license extension; license transfers; etc. Since the program proposed in this paper would focus on new plant capital costs, resolution of all of the current plant issues will be necessary to minimize the uncertainties in total life cycle costs for new plants.

CONCLUSION

Successful completion of the program described above would result in simpler, economically competitive nuclear plant designs that could be deployed in the U.S. and overseas. Along with those designs would be a new regulatory infrastructure at NRC that is more risk-based and less prescriptive.

Such a program would provide a single clear unifying theme for government and industry sponsored nuclear R&D programs for the next decade. It would encourage, if not require, that government, industry, national labs, and universities work together toward a common goal. Involvement in such an undertaking could also help to provide the means for converting some of the national labs' activities from defense projects to commercial projects. Involvement by the universities would encourage students to consider the nuclear energy career option.

The potential for near term application to operating plants in the U.S. and to near term deployment of new plants overseas could be very substantial. Overseas organizations should be encouraged to participate in the program. These near term applications would also provide important feedback, so that the R&D program can be regularly adjusted to assure that the end products will be commercially viable in an ever-changing competitive marketplace.

WORKING TOWARDS PUBLIC ACCEPTANCE IN GERMANY

Dr. Peter Haug

Managing Director
German Atomic Forum
Heussallee 10
D-53113 Bonn

(VIEWGRAPH 1: WORKING TOWARDS PUBLIC ACCEPTANCE IN GERMANY

Ladies and Gentlemen,

the organizers of this conference invited me to talk to you about "Working towards public acceptance in Germany: Past achievements and present challenges".

My presentation will be divided into four main parts:
- First, I will talk about the conceptual and strategical ideas behind the work of our public relations initiative, "Informationskreis Kernenergie", or "Nuclear Information Group".
- In a second part, I will very shortly present an overview of our main public relations activities.
- Thirdly, I will make some comments on the present situation in Germany, after the federal elections, and
- Finally, I will try to very shortly extract what I think we have learned from more than 20 years of experience in this field.

My presentation will keep to the following outline:

(VIEWGRAPH 2: OUTLINE 1)

1. INTRODUCTION

Let me begin by giving you a personal credo: It is my deep conviction, that public acceptance of nuclear power in any specific country is very strongly influenced by the structure of this country's political, its social and its media system, its history and life-style. Acceptance may even vary strongly within a country from one region to another.

Working toward Public Acceptance in Germany

Outline

1. Introduction
2. Nuclear Energy in Germany: The Facts
3. Public Acceptance and Public Relations Work
4. "Informationskreis Kernenergie": Communicating for Nuclear
5. Strategial Issues of "IK"
6. Defining a PR-Programm
7. Public Relations Actions of "IK"
8. Present challenges after the elections
9. Lessons Learned from IK Activities

German Utilities '98

Scope	Main Activity	Number of Utilities
nationwide	production and distribution	8
regional	distribution and production	30
local	distribution	680
TOTAL		718

Nuclear Energy in Germany '97

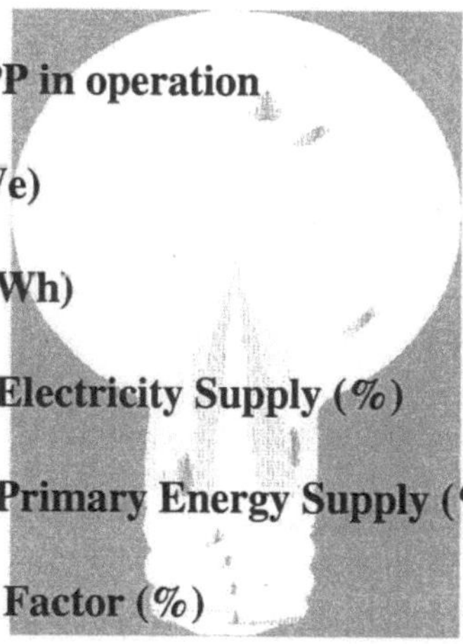

Number of NPP in operation	19
Capacity (MWe)	22'149
Production (TWh)	170.4
Percentage of Electricity Supply (%)	36
Percentage of Primary Energy Supply (%)	13
Average Load Factor (%)	92

World Championship in Electricity Production per Station and Year

1980	NPP Unterweser	9.81 Billion kWh (World record)
1981	NPP Unterweser	9.54 Billion kWh
1982	NPP Biblis B	9.74 Billion kWh
1983	NPP Grafenrheinfeld	9.96 Billion kWh (World record)
1984	NPP Grafenrheinfeld	10.15 Billion kWh (World record)
1985	NPP Grohnde	11.48 Billion kWh (World record)
1986	NPP Grohnde	10.79 Billion kWh
1987	NPP Grohnde	10.21 Billion kWh
1988	NPP Palo Verde-3 (USA)	10.86 Billion kWh
1989	NPP Grohnde	10.86 Billion kWh
1990	NPP Grohnde	10.69 Billion kWh
1991	NPP Emsland	10.83 Billion kWh
1992	NPP Brokdorf	11.33 Billion kWh
1993	NPP Unterweser	11.40 Billion kWh
1994	NPP Isar-2	11.13 Billion kWh
1995	NPP Grohnde	11.36 Billion kWh
1996	Philippsburg 2	11.47 Billion kWh
1997	Grohnde	12.53 Billion kWh (World record)

Top Ten

International ranking, based on electricity production per station and year

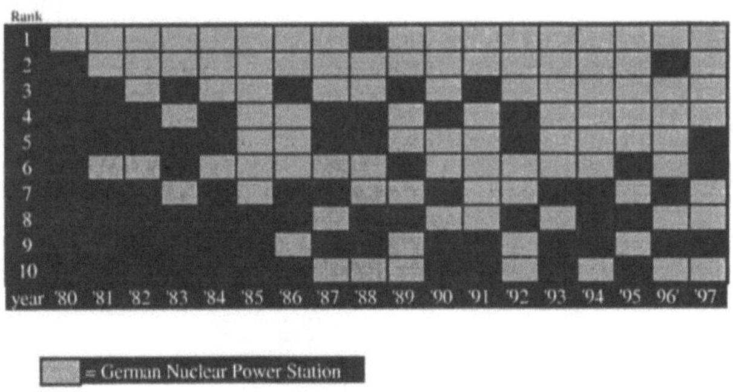

Germany is a typical example of a highly decentralized system in political decision making and also in electricity supply.

My presentation, therefore, will describe what we learned from working in such a system. Experiences in different countries may vary considerably, given a possibly completely different structure there.

2. NUCLEAR ENERGY IN GERMANY: THE FACTS

Now, what is so special about the German situation?

(VIEWGRAPH 3: GERMAN UTILITIES '98)

In my country, we count more than 700 utilities, from very small ones to very big and successful units. These utilities show a typical 3 level structure: We have local distributors, regional producers/distributors and 8 nation-wide producers and distributors. Only the latter have the necessary financial and organizational means for running nuclear power stations.

(VIEWGRAPH 4: NUCLEAR ENERGY IN GERMANY '97)

In 1997, a total of 19 commercial nuclear power plants, all located in Western Germany, with a capacity of 22'149 MWe, produced approximately 170,4 TWh. Nuclear energy thus made up for about 13 % of overall energy consumption and for more than one third of electricity demand. Excellent time availability figures and load factors have made German nuclear power stations rank among the world's leading plants for many years in a row.

(VIEWGRAPH 5: WORLD CHAMPIONSHIP IN ELECTRICITY)
(VIEWGRAPH 6: TOP TEN)

3. PUBLIC ACCEPTANCE AND PUBLIC RELATIONS WORK

Quite contrary to this success story, public acceptance dropped sharply after the Tchernobyl accident from its previous, relatively small, but stable majority of about 40 percent pros, 35 percent cons and 25 percent "do not knows". The latest survey, carried out in 1998, shows us that again there has been a distinct positive change of opinion with regard to nuclear energy in recent years.

A trend analysis of the period between 1993 and today shows that the willingness to accept investments in replacing old nuclear facilities, or even to accept the creation of new facilities, has again increased considerably. In early 1993, 27 percent of the West German population favored investing in replacement of nuclear facilities, as compared with 40 percent currently; in a parallel development, the share of Germans who advocate expanding the use of nuclear has doubled, increasing from 3 to 6 percent. During the same period in West Germany, the share of persons committed to phasing out nuclear energy declined, from 32 percent to 24 percent. In East Germany, where attitudes towards nuclear energy were more positive initially than in West Germany, public opinion has essentially remained stable.

Members

Utility members

Bayernwerk AG
Energie Baden-Württemberg
Hamburgische Electrizitätswerke
Isar-Amperwerke AG
Neckarwerke Stuttgart
PreussenElektra AG
RWE Energie AG
Vereinigte Elektrizitätswerke Westfalen

Suppliers members

Asea Brown Boverie
Siemens/Kraftwerkunion

Consulting members

GRS: Reactor Safety Research
GFS: Radiation Protection
KfK: Nuclear Research Center
KFA: Nuclear Research Center
DBE: Waste Management
BfS: Radiation Protection

Basic Principles of Work

Defining a PR-Program

- **Analyze the public acceptance situation.**
- **Work out a communication strategy, defining**

  - the target groups you want to reach,
  - the topics you want to communicate,
  - the goals you want to reach with your communication.

- **Select appropriate proffessional support, i.e. PR-agencies, etc.**

Target Groups

Our main target groups are decision makers and decision multipliers, including the young generation, from

- **Politics (federal, regional, local)**
- **Media (print, electronic)**
- **Main social groups (labour unions, churches, police, military, industry associations)**
- **Insiders (i.e. employees from utilities and vendors)**

4. "INFORMATIONSKREIS KERNENERGIE (IK)": COMMUNICATING FOR NUCLEAR

Existing information centres at nuclear power plant sites very well cover local information needs. Regional utilities cope with regional needs. But very soon in the beginning of the nuclear controversy in our country it became obvious, that it was necessary to establish an additional, strictly federal, which means central, organization that can take care of all nation-wide information needs.

Therefore, major German utilities and vendors founded what we call today "Informationskreis Kernenergie" as a common public relations organization for nuclear. It combines the intellectual and financial means of its founding members and has been active for more than twenty years now.

(VIEWGRAPH 7: MEMBERS)

5. STRATEGICAL ISSUES OF "IK"

The main object of our work is to try to convince the public that nuclear energy is a safe, economic and environmentally tolerable source of primary energy. Our work is being backed by nuclear and public relations specialists from vendors, utilities, research centres and government organizations.

It is very important in our work to make sure that we develop and take all actions in close cooperation and in periodical consultation with the founding members of IK.

(VIEWGRAPH 8: BASIC PRINCIPLES OF WORK)

The main principles of our work are: <u>independence</u> from political parties and individual companies; restriction to <u>non-local</u> public relations work (because the local work is done by the Nuclear Power stations themselves!); and <u>limitation</u> to selected target groups.

6. DEFINING A PR PROGRAM

How do we proceed in order to arrive at a detailed annual program for our work?

(VIEWGRAPH 9: DEFINING A PR-PROGRAM)

There are actually three steps to go through: First, we do an indepth analysis of the public acceptance situation. This will give us a first impression and point out possible problems, either locally or topically. Then we work out a communication strategy which builds upon this analysis. It should at least define the most promising target groups for our communication, make some clear statements about the topics we want to address and also define the goals we want to reach with our communication efforts. And finally, we always look for professional support - either for conceptual details or for carrying out the final PR-program. At present, we are addressing the following <u>target groups</u>:

Topics

- **Importance of nuclear energy for Germany as an industrial site**
- **Management of radioactive waste**
- **Nuclear energy and climate change**
- **Development of advanced reactors with improved safety features**
- **Special safety advantages of German reactors**

Our Communication Goals

We want to communicate the following messages to our target groups:

- **Electricity production in nuclear power plants is safe.**
- **Nuclear energy has many ecological advantages (no $C0_2$ emission).**
- **Nuclear energy cannot be given up in view of increasing world energy needs.**
- **Nuclear energy makes sense economically and ecologically.**
- **Nuclear energy saves resources and gives economic independence.**

Booklet "Nuclear Energy: Basic Knowledge"

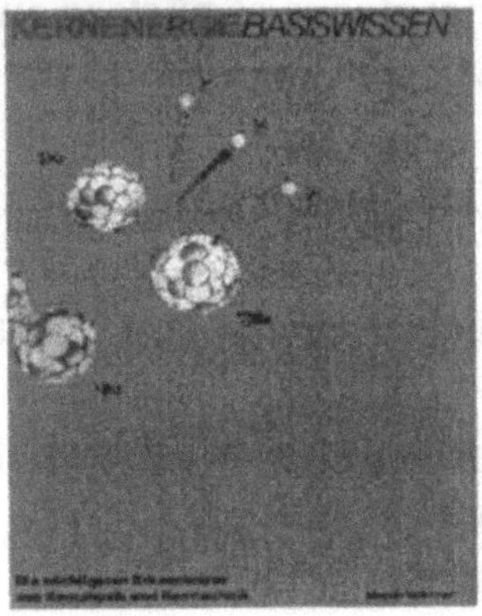

- total number printed: 1.150.000 copies
- distributed in 1997: 45.938 copies

Booklet "Radioactivity and Radiation Protection"

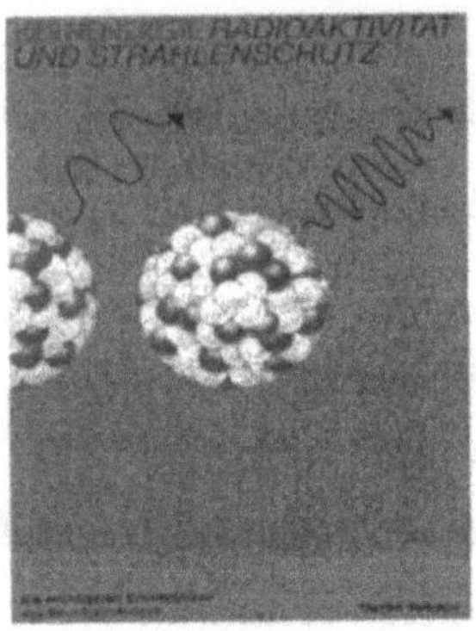

- total number printed: 220.000 copies
- distributed in 1997: 30.258 copies

Booklet "Nuclear Energy: clear and simple"

- total number printed: 600.000 copies
- distributed in 1997: 12.005 copies

Booklet "Chernobyl: The Accident"

- a documentation of todays knowledge about the accident, its causes, its effects - locally and internationally
- total number printed: 80.000 copies
- distributed in 1997: 13.363 copies

Book "Nuclear Energy in Germany"

- Total number printed:
  270.000 copies
- distributed in 1997:
  13.977 copies

Poster "Nuclear Energy Power Stations"

- total number printed:
  545.000 copies
- distributed in 1997:
  20.715 copies

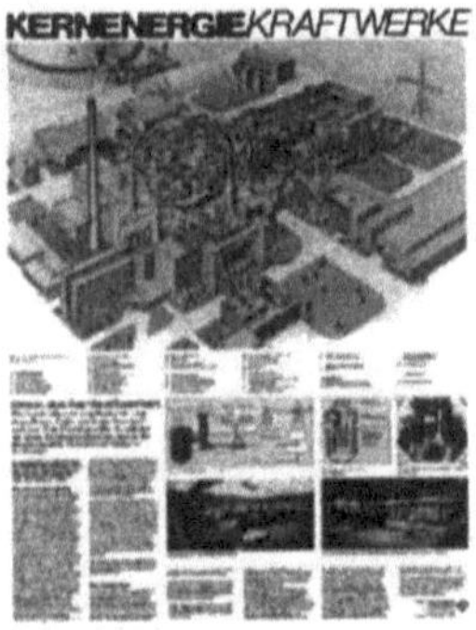

Poster "Nuclear Energy: Sites"

- total number printed: 1.270.000 copies
- distributed in 1997: 10.393 copies

Illustrations service for newspapers

- in 1997, a total of 6 illustrations was sent to about 1.600 newspapers producers, each.

Transparencies

- a total of 2.000 copies was produced for special interest groups
- distributed in 1997: 669 copies

Information Service in MS-DOS Format

- total number produced in 1997:
  93.000 copies
- contains a complete data base of nuclear related topics

Adventure Game on CD-ROM

- learning nuclear by playing an exciting and funny game
- knowledge must be acquired in the course of the game
- Typical "infotainment" product for young people
- total number produced in 1997:

  5.000 copies

Internet Homepage and Services

- Started 1.1.1997
- contains a very substantial encyclopedia on nuclear energy
- interesting services for young people and students
- links are given to other nuclear related homepages of the Internet

Movie "Transport of used fuel elements: a documentation"

- story-board and treatment completed in 1996
- production was realized in 1997

Available Movies on Nuclear Energy

- Continuous offer of 5 different movies, which are being distributed with 20 copies, each
- One booklet, which lists all nuclear related movies, giving the producer, the year of production and the organization, where one can get it from

(VIEWGRAPH 10: TARGET GROUPS)

The topics we have selected for our present campaign are given in the next chart.

(VIEWGRAPH 11: TOPICS)

And finally, the goals of our communication program are written down on the next chart.

(VIEWGRAPH 12: OUR COMMUNICATION GOALS)

Let me stop now with the theory and let us turn to a very short look at practical examples of our work.

7. PUBLIC RELATIONS ACTIONS OF "IK"

PRINTED MEDIA/BASE LOAD MEDIA:

Our "Basic Load Media" are publications that were developed in past years and have proven themselves in every day use. They will be kept up to date and reprinted, if demand keeps on.

(VIEWGRAPH 13: BOOKLET "NUCLEAR ENERGY: BASIC KNOWLEDGE)
(VIEWGRAPH 14: BOOKLET "RADIOACTIVITY AND RADIATION PROTECTION")
(VIEWGRAPH 15: BOOKLET "NUCLEAR ENERGY: CLEAR AND SIMPLE")
(VIEWGRAPH 16: BOOKLET "CHERNOBYL: THE ACCIDENT")
(VIEWGRAPH 17: PICTUREBOOK "NUCLEAR ENERGY IN GERMANY")
(VIEWGRAPH 18: POSTER "NUCLEAR ENERGY: POWER STATIONS")
(VIEWGRAPH 19: POSTER "NUCLEAR ENERGY: SITES")

AUDIOVISUAL MEDIA:

The main advantage of "Audiovisual Media" is, of course, their intensive information transport and lasting perception. Complicated facts, which would be difficult to explain otherwise, can be clarified easily. We offer a full scope of transparencies, computer programs and videos.

(VIEWGRAPH 20: ILLUSTRATIONS SERVICE FOR NEWSPAPERS)
(VIEWGRAPH 21: TRANSPARENCIES)
(VIEWGRAPH 22: INFORMATION SERVICE IN MS-DOS-FORMAT)
(VIEWGRAPH 23: ADVENTURE GAME ON CD-ROM)
(VIEWGRAPH 24: INTERNET HOMEPAGE AND SERVICES)
(VIEWGRAPH 25: MOVIE: "TRANSPORT OF USED FUEL ELEMENTS: A DOCUMENTATION")
(VIEWGRAPH 26: AVAILABLE MOVIES ON NUCLEAR ENERGY)

Advertisement Series: Nuclear News

- Approximately 24 different, short news have been published in 1997 in the following important German newspapers: Frankfurter Allgemeine Zeitung, Die Welt, Süddeutsche Zeitung, Handelsblatt, Die Zeit, Welt am Sonntag, Focus, Spiegel, Wirtschaftswoche

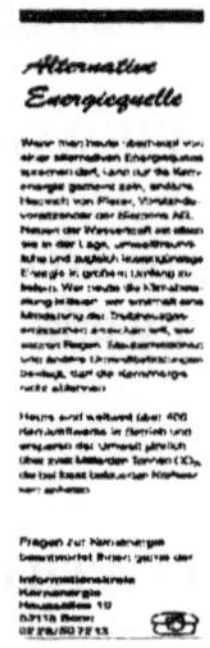

Information Trips for Journalists

- 13./14.2.97 COGEMA/La Hague
- 8./9.4.97 COGEMA/La Hague
- 3./4.6.97 Gorleben
- 2./3.9.97 Ahaus
- Excellent reports, even from critical journalists

Information Center "Forum in the Press House"

- targeted at MP's and employees of federal ministries
- very high level communication platform
- 19 presentations in 1996, with 100 high level decision makers, each
- 37 background discussions with journalists on selected nuclear topics
- 3 art exhibitions

Opinion Research

- Continuous monitoring of the public's opinion on nuclear energy
- Thorough evaluation of all opinion research results

Answering Service

- Individual answers to about 15.828 letters and postcards were organized in 1997
- The topics of main interest are:
  - waste management, final storage
  - safety of nuclear stations
  - advanced reactor designs

Our Bestsellers

Circulation	
545.000	Poster "Nuclear Energy: Power Stations"
1.150.000	Booklet "Nuclear Energy: Basic Knowledge"
1.270.000	Poster "Nuclear Energy: Sites"
270.000	Pictorial Guide "Nuclear Energy in Germany"
550.000	Booklet "Nuclear Energy: clear and simple"
270.000	Booklet "Radioactivity and Radiation Protection"

Results of German Federal Elections September 27th, 1998

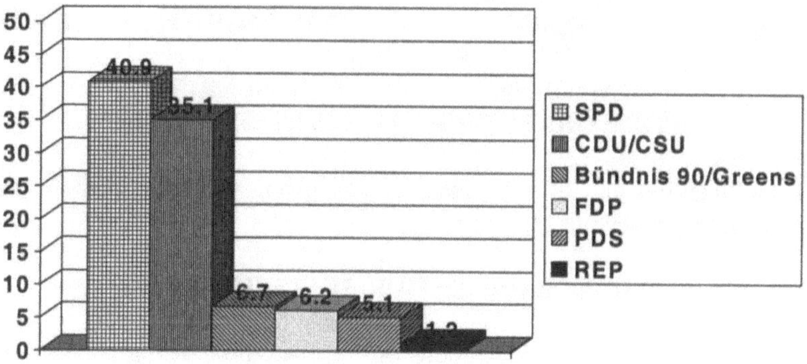

Seats: Total Seats in the Lower House of the Federal Parliament

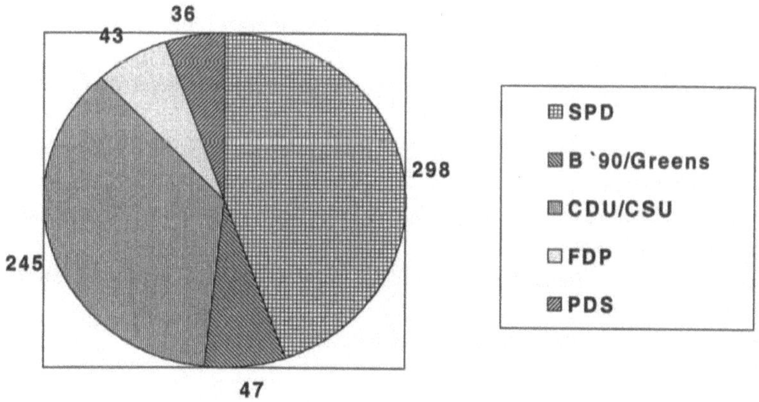

SHORT TERM INFORMATION:

Since nuclear energy is mainly discussed in a public debate, it is important to have available detailed information quickly and accurately.

The media in this group give us the possibility to pass on information speedily and/or to react upon current events quickly.

(VIEWGRAPH 27: ADVERTISEMENT SERIES: NUCLEAR NEWS)
(VIEWGRAPH 28: INFORMATION TRIPS FOR JOURNALISTS
(VIEWGRAPH 29: INFORMATION CENTER: "FORUM IN THE PRESS HOUSE")
(VIEWGRAPH 30: OPINION RESEARCH)
(VIEWGRAPH 31: ANSWERING SERVICE)

Let me finish this part of my presentation by giving you some additional figures, which are supposed to demonstrate the range of our efforts.

(VIEWGRAPH 32: OUR BESTSELLERS)

In this chart, you find the circulations of our main, best selling publications as of the end of 1997. Some of them have really gone very far into our population.

8. PRESENT CHALLENGES: AFTER THE ELECTIONS

Let us turn abruptly to recent devellopments: The recent federal elections very probably ended the 16 years' of chancellorship for Dr. Helmut Kohl, the conservative party:

(VIEWGRAPH 33: RESULTS OF GERMAN FEDERAL ELECTIONS)

- The majority of votes went to the Social Democrates under Gerhard Schröder, making 40,9 %.
- The Christian Democratic Union of Helmut Kohl and his closest party allies reached only 35,1 %.
- The Greens lost approximately 2 %, but still arrived to be the third most important political power with 6,7 %.
- The Liberal party, the former coalition partner of Kohl, arrived at a remarkable 6,2 %, which, however, was not enough to compensate ist partners' losses.
- Amazingly enough, the former communists, now called PDS, came out shortly above the 5 % cut-off limit.
- All right wing parties together stayed below 2 %.

These results give a rather stable majority of a red-green Government.

(VIEWGRAPH 34: SEATS)

Lessons Learned from "Informationskreis Kernenergie" Activities

- **Identify your allies in every possible area of common interest and try to get them involved.**
- **Identify your most important target groups at all levels and decide which organization(s) and/or persons should concentrate on them.**
- **Try to get political assistance at all levels and at a very early state.**
- **Make it clear to utility leaders and/or constructors that an extensive PR-program with an adequate funding is essential right from the very first moment of planing for a nuclear site.**
- **Get assistance from the very best PR-professionals available in your country.**
- **Make sure your final program will cover local as well as regional as nationwide PR-needs.**

What does this mean for nuclear energy in Germany? Although it is to early to know too many details, we have to get prepared for the following political plan of action:

1. The official policy of the new Government will be to phase out of nuclear "as soon as possible". The Greens intend to spend about 4 years, the Socialists about 20-40 years in order to reach this goal.

2. The waste management question will be reopened from the start. Reprocessing might be forbidden altogether. The Gorleben repository might be declared "unsuited" – which, of course, is not the case. The Morsleben repository will be closed; the Konrad project will be ended.

3. In order to avoid court trials with liability sues of up to 200 billion DM, Mr. Schröder "invites" representatives from the German utilities to talks in an effort to arrive at negoticated, consentual results. The time frame set for negociations is just one year. Nobody knows, what the results of these negociations will be. And of course, nobody can tell, what might happen if a consensus is not reached.
   So be prepared to experience an interesting year of nuclear discussions in Germany!

Now, finally, I will close my presentation by extracting some recommendations from our work.

(VIEWGRAPH 35: LESSONS LEARNED FROM "INFORMATIONSKREIS KERNENERGIE" ACTIVITIES)

9. LESSONS LEARNED FROM "INFORMATIONSKREIS KERNENERGIE" ACTIVITIES

Let me try to summarize our past experience in the following recommendations:

1. Identify your allies in every possible area of common interest and try to get them involved.

2. Identify your most important target groups at all levels and decide which organization(s) and/or persons should concentrate on them.

3. Try to get political assistance at all levels and at a very early state.

4. Make it clear to utility leaders and/or constructors that an extensive PR-program with an adequate funding is essential right from the very first moment of planning for a nuclear site.

5. Get assistance from the very best PR-professionals available in your country.

6. Make sure your final program will cover local as well as regional as nationwide PR-needs.

Thank you very much for your interest!

Public and Policymaker Support for Nuclear Energy

Angelina S. Howard

Senior Vice President
Industry Communications
Nuclear Energy Institute
Washington, DC 20006

Good afternoon.

I am delighted to be here today in the "The City of Lights" to talk about public and policymaker support for nuclear energy in the United States.

These are very exciting times for the U.S. nuclear energy industry. We are witnessing the convergence of several major policy issues that are closely linked to our industry's future. Issues like the environment, competition among electricity generators, reform of the regulatory process, disposal of used nuclear fuel and public acceptance of nuclear energy.

All these issues are moving in a way that is positive for nuclear energy—and for the United States. What is even *more* exciting is that the same thing appears to be happening all over the world.

At the Nuclear Energy Institute's annual conference last May, we unveiled a new strategic direction for the 21st century titled, **Nuclear Energy: 2000 and Beyond**. This strategic direction outlines key policy priorities for the industry. It provides a framework for conducting a reasoned discussion on how to derive the greatest benefit from nuclear energy. There are copies of the document available at the back of the room, and I encourage all of those interested to take home a copy.

Today, I am going to address one element of the new strategy: The clear link between these converging issues and positive changes in perceptions about nuclear energy as we move toward the next century.

There is clear evidence among U.S. policymakers that *their* perceptions about nuclear energy's role in our nation's energy portfolio are changing for the better. Support for nuclear energy in the administration and in Congress is growing.

By way of example, 70 percent of Congress voted in favor of a measure to reform the used fuel management program even though a political maneuver kept it from becoming law. In the Senate, Chairman of the Appropriations Committee's subcommittee on energy and water, Pete Domenici terms nuclear energy "essential to the continued prosperity" of the United States, and says the administration "needs to stop hedging on this issue." The support among policymakers is not unanimous, of course, but a dramatic

Preparing the Ground for Renewal of Nuclear Power, edited by
Kursunoglu *et al.* Kluwer Academic / Plenum Publishers, New York 1999.

turnaround is taking place. And it is the positive actions being proposed and taken by policymakers that ultimately will influence the media and the public.

We already are seeing among the American public very positive signs, as well. A recent national survey looked at the views of an influential sector of the American public—college graduates who are registered to vote. Nearly nine out of ten agree that we should renew the licenses of nuclear plants that continue to meet safety standards. Three out of four agree that we should keep our existing nuclear plants and that we should keep open the option to build more nuclear power plants in the future.

More to the point, when asked which source of electricity will be used most in the United States 15 years from now, nuclear energy was mentioned by more people than any other source—45 percent.

The survey also uncovered additional points that bear close examination. For example, people do not know enough about what the nuclear industry does and how we do it. In fact, members of the public are surprised to learn that there are more than 100 nuclear units generating 20 percent of U.S. electricity. Most would guess less than 10 plants are in operation. Nor does the public understand the science of nuclear energy, or the lengths to which the industry goes to provide safe, reliable and economical electricity. And the public is just coming to understand the beneficial environmental impact of nuclear energy. So while nuclear energy and technologies are becoming more accepted, an image problem persists even among the aware public.

A companion survey of the general American public this summer also indicated majority support for nuclear energy, for license renewal, for keeping our existing plants and for keeping the option to build more plants in the future. Although the support was not as strong as with the college-educated voters, the differences were surprisingly small.

A not surprising finding common to all three surveys was that favorable attitudes about nuclear energy *increased* when respondents heard information about its clean air benefits. Among the college-educated voters, the percentage who favored nuclear energy jumped from 65 percent to 75 percent after they learned that there are more than 100 nuclear plants in the United States that generate one-fifth of the electricity we use without emitting any greenhouse gases. The percentage who *strongly* opposed nuclear energy was cut nearly in half—from 13 percent to 7 percent. Among the general public, there was a similar shift toward favorability—especially among women.

Another finding from the survey of college graduates is that those who feel more informed are more favorable toward nuclear energy. The survey showed:

- a 78 percent approval rating among those who described themselves as *very well* informed
- a 67 percent approval rating among those who described themselves as *well* informed, and
- a 60 percent approval rating among those who described themselves as *not* well informed.

There is a valuable lesson here: No one benefits when the industry fails to convey good news. The industry will miss an extraordinary opportunity to make a positive impact on public opinion if it just remains quietly on the sidelines, bemoaning the lack of support for this vital, emission-free energy source.

It is time to abandon the passive "no news is good news" approach to communication.

It is time to aggressively get the word out about our true value and our real achievements.

It is time to stop allowing the small but vocal anti-nuclear interest groups to define the public debate and unduly influence public energy policy. We have a lot of great news to share and we need not be apologetic about that.

- Ours is a non-polluting, zero-emissions technology that can help electricity generators around the world meet stricter air quality standards.
- Ours is an economically competitive industry that—on a level regulatory playing field—can hold its own with, and even surpass, other generating sources.
- Our industry has posted a steady decade-and-a-half of improvement in operating efficiency and output—with no sacrifices to safety or cost control.

The industry needs to disseminate this good news aggressively, not just in the United States, but in other countries as well. By this I mean nuclear operators, not just industry associations. While NEI considers building public understanding of nuclear energy's benefits to be one of its most vital missions, we also have to have the message going out locally from the companies operating these plants. To support its members and the industry, NEI carries out a comprehensive communications program on the benefits of nuclear energy and technology.

Being here in France, I would be remiss if I failed to salute EdF's stellar and long-term effort to educate the French public about the importance of nuclear energy. Its numerous and informative advertising campaigns over the years are just part of the reason why it is held in such high public esteem.

As an industry, we have to follow that example and use the news media to reshape public and policymaker thinking. These media allow us to demystify the technology and clarify the benefits of nuclear energy for a broad audience.

As already noted, a major element of our research this year dealt with *perceptions*. While two-thirds of those polled said they favored the use of nuclear energy, only one-fifth felt that the majority of their neighbors were favorable to nuclear power. *Two-thirds* were favorable toward nuclear energy, but only *one-fifth* felt that a majority of their neighbors were also supporters. Yet the survey of the general public conducted at the same time showed that a clear majority favored nuclear energy.

The same perception gap is found among legislators and policymakers. Although a large majority of policymakers support nuclear energy, they perceive that their constituents do not. This constrains their willingness to exhibit the strong leadership necessary to support key nuclear energy initiatives. NEI is working vigorously to close that gap.

Earlier this month, NEI conducted two focus groups to see if we could learn *why* supporters of nuclear energy think their neighbors believe otherwise. No definitive answer emerged, but we did discover several clues.

Some focus group participants perceived the public at large as fearful of nuclear energy because of its potential dangers based on lingering images of Hiroshima and Chernobyl. Some pointed to the news media's relentless coverage of the negative. While many believed *they* were independent enough to form a balanced opinion of nuclear energy, they believed *others* were not.

Once the group learned that the majority of the public does indeed favor nuclear energy, they were surprised. Many simply assumed that since *their* positive perceptions of nuclear energy had not changed over time, their neighbor's negative perceptions about nuclear energy had not changed either. But people have a tendency to forget even negative images over time, and nuclear energy has not been a conspicuous issue recently. Another explanation was that opponents are more outspoken and passionate than supporters of nuclear energy.

The focus groups also offered insight on how to bridge the perception gap. They suggested that people form impressions of public opinion based on what they see and hear—such as negative news images—as well as on what they *do not* see and hear—namely, positive coverage of nuclear energy on a routine basis.

Their comments emphasized what research has shown for years—that support for nuclear energy is driven primarily by perceptions of the need for nuclear energy and its benefits. The participants knew the benefits, but believed most people did not. If they saw more information in the media about the benefits of nuclear energy, they would be more inclined to believe that other Americans shared their opinion because they were similarly exposed to the information.

That would help narrow the perception gap. And that's exactly what NEI hopes to see happen as a result of our focused efforts to communicate the benefits of nuclear technologies. The perception gap must be closed because policymaker and community support of our business are essential. The industry has the support for nuclear energy it needs in the United States. It simply needs for it to be brought out into the open and recognized.

As events in Germany attest, work needs to be done here in Europe, as well as in the United States, to convey the environmental benefits of nuclear energy. The industry must continue to build understanding of its essential role as an emission-free energy source. Experience, and our research, suggest that the industry *can* redefine the public's perceptions and influence their actions. By skillfully using communications resources, we can redefine the public debate on more favorable terms.

The industry must build understanding of its essential role as an emission-free energy source. Experience, and our research, suggest that the industry *can* redefine the public's perceptions and influence their actions. By skillfully using communications resources, we can redefine the public debate on more favorable terms.

The public is receptive to new messages, and the industry has the ability to make new messages work. With new awareness and a more positive attitude, we will increase public and policymaker support for nuclear energy. We will also achieve much needed recognition of that support. And we will help that majority of nuclear energy supporters feel secure and positive about why they favor the continued support for nuclear energy. All are essential if the benefits of nuclear energy are to be fully enjoyed by future generations.

Thank you.

GLOBAL FOUNDATION CONFERENCE

Session III - Social Acceptance

Annotator: Jacques de la Ferté

Head, External Relations and Public Affairs
OECD Nuclear Energy Agency (OECD/NEA)
Le Seine St Germain
12, boulevard des Iles
92130 Issy-les-Moulineaux, France

It is a great privilege for me to be with you this afternoon in this session on "social acceptance". The last three speakers have skilfully illustrated specific aspects of this complex subject and I do not intend to challenge them on their respective grounds.

Rather I take it that, as an annotator, I should be permitted to highlight two or three facets of this incrasingly important social acceptance concept, which has become a sine qua non of the further development of the technology in general, and nuclear energy in particular.

Few modern industries are as closely dependent on non-technical and social acceptance factors as the nuclear industry.

Many historic and certain ever-lasting reasons can be recalled. But, more important is the fact that, in the last twenty years, the context of societal decision-making has changed radically. Superficially, the old forms of decision-making remain unchanged. Traditionally, elections remain the general process by which citizen delegate their powers to a few elitist circles, which will take the decision on their behalf. But today, the role of the citizen is no longer confined to that of legitimating the members of these circles.

Decision-making routes have taken new forms. The main reasons are:

- the increasing influence of environmental values;
- the rising education levels in the public;
- the influence of the media and opinion polls.

The most motivated segments of the public have gained new powers, and a self-attributed legitimacy that is stronger than that of their elected representatives. Therefore, the questions now is not "how to circumvent public opposition", but how to adapt to the conflictual reality that characterises many decision processes at national and local levels.

In other words, how to steer the decision-making process along a gradual negotiation pattern leading to a "modus vivendi" or a "negotiated consensus" that will satisfy as many stakeholders as possible, that the solution reached will be technically, economically and socially acceptable.

Preparing the Ground for Renewal of Nuclear Power, edited by
Kursunoglu *et al.* Kluwer Academic / Plenum Publishers, New York 1999.

Nuclear energy decision-making is perhaps the most dramatic illustration of this trend. Past failures and current experience with nuclear plant and radioactive waste management site selection have brought forward a number of points. First remark, communication is an indispensable tool to identify, assemble, confront, reconcile the various pieces of information and opinions upon which a consensus can be built. In particular, a two-way communication process will allow the scientific-technical-economic approach to interact with peoples' values and symbols, and eventually find a point of equilibrium. But, of course, the public is not a single entity. So communication must be tailor-made.

Second remark, the complexity of the social acceptability process is compounded by the fact that it incorporates a wide range of factors, some with an objective, measurable value, others with a subjective, sometimes emotional contents, the value of which cannot be measured accurately.

Taking into account subjective values of stakeholders or the population in general, raises a true society issue, indeed a dilemma, which ultimately can only be solved at the political level.

The issue includes two different questions: first, which level of risk are we prepared to accept for our society? Zero risk does not exist, but 100% protection against any risk is not achievable either. Second, to what extent should erroneous perceptions be taken into account in the decision to use a technology, or to shelve it?

How do we handle the case of a technology like nuclear in a world where public feelings include a part of ignorance, manipulation and politics? Today the degree of social acceptance of nuclear plant siting, operational safety, and particularly radioactive waste management, will ultimately decide on the overall future of nuclear energy.

Actually I should have said "the degree of social and political acceptance ...". Indeed, the fact that nuclear energy has become an important theme in political programmes and in regional and national elections contributes to further dissociate the perception of this form of energy from the realities of the nuclear technology.

This is to be regretted, and it will remain quite a challenge to achieve an objective and dispassionate debate on the subject.

Nevertheless, I believe there are two important routes, though not really new, which need continuous follow-up. One is to strengthen the teaching of young generations, at school, on how to form a rational, independent judgement on societal issues, and their underlying technological options, and the other is to continue to demonstrate that nuclear power is a reality, and publicise the fact that nuclear plants work, produce electricity for the people, do not produce GHG, and create jobs.

PUTTING BALANCE INTO SAFETY REGULATION

Richard Wilson

Mallinckrodt Professor of Physics
Physics Department
Harvard University
Cambridge, MA 02138

INTRODUCTION

In the USA we now emit 11% more CO_2 than in 1990; and at Kyoto we promised to reduce CO_2 emissions to 8% below 1990 levels in 10 years for a decrease of 19% below today's levels. If all the electricity now generated by nuclear power were to be generated by coal that would increase CO_2 another 8% making it more difficult to meet our commitment if we abandon nuclear power. My purpose today is to explain to you that we COULD meet this commitment, or nearly so by a nationwide, and of course hopefully worldwide decision to use nuclear energy. I will also explain why it should it be too expensive.

I do not want to belittle the difficulties in replacing the burning of coal with nuclear power. These difficulties are NOT technical but social and political. In spite of the rhetoric of the politicians, I know of no one who believes that the USA *will* meet its commitments under Kyoto, except perhaps by an accounting trick.[1] We will continue to burn coal - probably in increasing amounts. As an example of this (Figure 1)[2] shows the railroad built by Burlington Northern 20 years ago (and jointly operated with the Union Pacific) to the Powder River basin in Wyoming. Originally a single track it is now double or triple track everywhere. This line has had up to 55 coal trains a day - each with 120 cars of 100 tons each. This area alone provides nearly 200 million tons a year - an amount which is expanding. Another railroad (Dakota, Minnesota and Eastern) is being built to join from the northeast in Minnesota with 300 miles of upgrade and 100 miles of new track. This would not be done if the coal industry had any doubts of their preeminence.

Moreover gasoline is now cheaper than it has ever been in real terms. In East Cambridge it can now be bought for 94 9/10 cents per gallon including taxes.

[1] This trick would be to count as a credit the amount of CO_2 that is soaked up each year by our regrowing forests. Of course this was also happening in 1990 but the letter of the Kyoto agreement does not demand that we take the 1990 carbon sequestering numbers into account.

[2] Photo by Mike Danneman

Preparing the Ground for Renewal of Nuclear Power, edited by
Kursunoglu *et al.* Kluwer Academic / Plenum Publishers, New York 1999.

Figure 1.

Contrary to what we all thought 25 years ago, natural gas seems to be plentiful and the (Carnot) efficiency of using it in combined cycle generators is now a phenomenal 58%.

Nonetheless I will try to paint a picture of what nuclear power could do. I will outline first what we know could be done technically and economically. Electricity from nuclear energy HAS been both safe and cheap. 25 years ago, Maine Yankee nuclear power plant had just been completed for $180 million, or $200 per KWe installed capacity. Connecticut Yankee nuclear power plant was producing electricity at 0.55 cents per kwh busbar cost, some part of which was paying off the mortgage. The operating cost was perhaps only 0.4 cents per kWh. Twenty-seven years ago, Benedict (1971) estimated average operating costs that were a little lower and capital costs about 25% higher than for Maine Yankee (Table 1). Taking no credit for learning, we could if we decide to return to the optimism and procedures of 25 years ago do as well. Allowing for inflation the operating cost could be less than 1 cent per kilowatt hour. Yet the average operating cost of nuclear plants in the USA today is 1.9 cents a kilowatt hour (McCoy, 1998) and for a well operating plant is still 1.4 (South Texas) to 1.5 (Seabrook) cents per kilowatt hour. These excludes such items as real estate taxes which can add another 0.5 cents/kwh. The construction cost of a new GE reactor is $1690 per kwe if built in 4 years, and more if it takes longer.

Table 1. Operating Costs

	Coal	Nuclear
Unit investment cost of plant, dollars/kw.	$202	$255
Annual capital charge rate per year	0.13	0.13
Kilowatt-hours generated per year per kw. capacity	5,256	5,256
Heat rate, million B.t.u./kwh.	0.009	0.0104
Cost of heat from fuel, cents/million B.t.u.	45	18
Cost of electricity, mills/kwh.:		
Plant investment	5.00	6.31
Operation and maintenance	0.30	0.38
Fuel	4.05	1.87
TOTAL	9.35	8.56

(Benedict, 1971)

It is useful to remember that one can choose almost any mixture of the following for a reactor:

Fuel: 3% enriched U, pure U235, Natural U, U233, Thorium, Pu239

Fuel Matrix: sintered oxide, metal, carbide, pebbles

Moderator: H_20,D_20, C, none

Heat transfer medium: H_20, Na, Pb/Bi, CO_2, He

Generator: Steam Turbine, Gas Turbine

The present Light Water Reactors use the first entry in each row and this choice was influenced by success of the naval reactors sponsored by Admiral Rickover - although the actual detailed designs are very different. It would take a billion or two dollars to bring an alternate type of reactor to the same level of development as the LWR although it is possible that one or another might eventually turn out to be a cheaper and/or safer choice. I will ignore this possibility for the principal purpose of this talk and explain how well we can do with present light water reactors.

But can we *increase* nuclear power generation fast enough to satisfy demand? Again we do not have to do better than in the past to be close to meeting the Kyoto agreement. In 1975 to 1985 five to ten nuclear power plants were being completed in the USA every year. In 1980 to 1990 five were being completed each year in France. The following countries NOW have the industrial capacity to build nuclear power plants and have built large fractions of them. USA, UK, Canada, France, Germany, Sweden, Russia, China, Korea, Japan, and India. Even allowing for no more countries to enter the industrial race the world could build 50 nuclear power plants a year. But the industry has wound down over the last 10 years and needs winding up. For example only one western country now makes large pressurized reactor vessels. This production rate could only be achieved after a delayed start of perhaps 5 years. In the USA we could have an additional 250 GWe by 2030. On this scenario, the US would miss the Kyoto deadline by about 10 years but be on the way to meeting future commitments.

Benedict (1971) reported that we had 20,000,000 tons of uranium at prices up to $100 per pound (Table 2) - and the higher prices only raised the operating cost by 0.5 cents a kilowatt hour - which would now be considered an acceptable increase. The quantity of uranium resources does not seem to have changed. The Uranium Institute reports that we have about 18 million metric tons of uranium in ore, proven reserves, reasonably assured supplies and possible supplies at prices up to $200 per kg. We can afford without appreciably increasing the initial fuel cost. This would produce in a light water reactor system about 4×10^{15} kwh of electricity, or enough for over a century at the postulated year 2030 demand of 2,500 kw.

Table 2. Uranium Supplies

Uranium price, $/lb.	U.S. uranium resources, tons	Increase in cost of electricity, mills/kwh.		Million kw.-years of electricity which could be generated	
		Water Reactor	Fast Breeder	Water Reactor	Fast Breeder
8	594,000	0.0	0.0	3,470	460,000
10	940,000	0.1	0.0	5,500	720,000
15	1,450,000	0.4	0.0	8,480	1,120,000
30	2,240,000	1.3	0.0	13,100	1,720,000
50	10,000,000	2.5	0.0	58,300	7,700,000
100	25,000,000	5.5	0.0	146,000	19,200,000

(Benedict, 1971)

There is a general rule about prices of any fuels. The time for depletion of reserves has stayed between 15 to 30 years for nearly a century! (Coal reserves are an exception). If there is enough fuel for 30 years there is little incentive for exploration but if the amount falls below 15 years the profit motive ensures that exploration restarts. The present and anticipated use of nuclear power provides no incentive to explore uranium. It appears we could have a future for nuclear power for 20 years without a breeder reactor, and possibly for many, many more. After perhaps half a century it would be wise to reconsider a breeder reactor using fast neutrons. Fast neutron reactors HAVE been built and DO work. The first nuclear powered electricity was from the EBRI reactor in Idaho when a whole town was lit by electricity as a public demonstration. But the cost, even after some experience and learning, is expected to be 20% higher than the cost of electricity from light water reactors and it will be at least 50 years before the increased cost of uranium compensates this. Nonetheless a breeder reactor should be competitive with other alternate electricity generating methods. If we have a breeder reactor system we can use all the U238 in the fuel and increase the effective fuel supply a factor of 100, we can use thorium which is 5 times as plentiful as uranium, and since the utilization of fuel is so much greater we can afford to use more expensive fuel. We can for example take the uranium from fly-ash from coal burning plants noting that there is more available energy from the uranium in coal than from the carbon. We might also extract the uranium from seawater. All in all, a factor of 1000 increase in fuel supply seems not unreasonable. It would be impudent to project the existence of the human race beyond the 100,000 years that this implies.

But we must bring the price down again (a factor of 2 in operating cost and 7 or so in construction costs). To do this it is useful to understand why the costs have gone up. I have seen NO careful study of this. Indeed in 1984 when the Energy Engineering Board of the US National Academy of Sciences proposed a study of the subject it was opposed by the utility industry, perhaps for fear of adversely influencing prudency hearings that were in progress before public utility commissions. Various ideas include the following:

- in 1970 manufacturers built turnkey plants or otherwise sold cheap reactors as loss leaders. But this can only account for a small proportion of the capital cost.
- Construction costs generally have risen since 1970 even when corrected for inflation.
- it may be that in 1972 we had good management and good technical people. But why has management got worse when that has not been true for other technologies?

Over Regulation

Most people seem to agree that the present limitation in nuclear power development in is public acceptance. This drives excessive regulation, and increases the cost. This increased cost often reaches a factor of three even after correction for inflation. It is probable that nuclear power plants are safer today than they were in 1972. But it would be hard to argue that the actual safety improvements that have occurred have been the cause of the threefold increase in cost. Most improvements are as a result of more careful thought using such approaches as event tree analysis, but without excessive hardware expense.

Many people have suggested that the problem is that the regulation is more than needed for adequate safety and this increases the cost (Towers and Perrin, 1995). In particular that regulation is too prescriptive and not based upon performance. A few of the arguments are roughly as follows:

The response to many regulations is to increase staff. The staff numbers at the Dresden power plant went from 250 in 1975 to over 1,300 today (Benhke, 1997). This costs money and it is far from clear that it improves safety.

The problem is not unique to the USA. In the UK the Atomic Energy Authority had to spend a lot of money making the Thorpe reprocessing plant as earthquake proof as an operating reactor - yet the inventory of dangerous material is far less (tens of Curies) and the danger of recriticality remote (Hill, 1997).

I want to address here the problem of regulation and the intricate and complex relationship between regulator and licensee. Although not an expert, I claim one advantage: I look on the problem from outside and I keep the three fundamental societal aims in mind. It is of course evident that safety is a large part of the cost of nuclear power. Without attention to safety a containment vessel, often 1/3 of the cost would not be necessary. But the important issue is how much safety should one pay for?

The first attempt to address "how much safety" for radiation exposure was begun by the first Nuclear Regulatory Commission (NRC) to take office some 14 years ago when astronaut William (Bill) Anders was chairman. After 2 years of public hearings started by the AEC the NRC set some radiation and safety guidelines. (NRC, 1976). The Commission proposed that expenditure on radiation exposure reduction should be made if it costs less than $1,000 per ManRem, now doubled to $2,000 (Kress, 1994) - a number higher than anyone in the hearings had proposed. Somewhat later the National Council for Radiation Protection and Measurements (NCRP, 1990) suggested that the number be between $10 and $100 per ManRem for dental and medical exposures. A corollary was implied but not explicitly stated. If a proposed dose-reducing action would cost more than this, it should NOT be done. I note that US EPA (1998) in their draft "Guidelines for Preparing Economic Analyses" suggest a number of about $5,800,000 per life saved which corresponds closely (using a linear dose response relationship) to the number in RM-30-2. I also note my view that that sum ($2,000 per manRem) should include the time spent talking about the particular issue.

In the 1980s the Advisory Committee on Reactor Safeguards (ACRS) made a study that led to the promulgation by the Commission in 1986 of a set of SAFETY GOALS (Federal Register, 1986).

"Individual members of the public should be provided a level of protection from the consequences of nuclear power plant operation such that individuals bear no significant additional risk to life and health" and

"Societal risks to life and health from nuclear power plant operation should be comparable to or less than the risks of generating electricity by viable competing technologies and should not be a significant addition to other risks".

These somewhat vague goals were supplemented by Quantitative Objectives:

"The risk to an average individual in the vicinity of a power plant of prompt fatalities that might result from reactor accidents should not exceed 0.1% of prompt fatality risks from the sum of prompt fatality risks from other accidents to which members of the U.S. population are generally exposed" and

"The risk to the population in the area near a nuclear power plant of cancer fatalities that might result from nuclear power plant operation should not exceed 0.15 of the sum of cancer fatality risks resulting from all other causes"

These objectives are met for light water reactors (with containments) is a subsidiary objective is met:

"The core melt frequency must be less than 1/10,000 per year"

Although not stated, it was implied that steps to decrease core melt frequency still further were unwarranted and it was not worth the expense to undertake them. For simplicity I will address this "intermediate" safety goal here but the same argument can be applied to the more fundamental safety goal.

There is a fundamental problem in implementing GOALS as opposed to issuing or following regulations. There is no definitive way of proceeding. But studies can be made retrospectively to see whether they are met. An independent study (Tengs, et al., 1995) suggests expenditures in the nuclear industry for RADWASTE have been over $1,000,000 per ManRem - 1000 times the goal. It seems that either the regulations (in this case probably the Technical specifications) are stricter than needed, that the industry is spending more than the regulations call for, or the total amount of money is so small it is not worth worrying about. The procedure does not, however suggest how they be relaxed or whether the cost decrease is large enough to be worth the bother. I suggest that the nuclear plant operators and the NRC, perhaps aided by IAEA since it is an international problem, should study the matter with some urgency.

Similarly the ACRS has repeatedly stated that it is not sensible to regulate on the basis of a Probabilistic Risk Assessment (PRA). But a PRA CAN be used, and in my view SHOULD be used to discuss retrospectively whether reactors that were designed and operate under existing regulations meet the goals. If they meet the goals, fine. If they do not regulations must be tightened. On the other hand if the safety goals are met with a large margin maybe the regulations can be relaxed. Indeed the important parts of a PRA can now be put on a small PC or laptop so that the effect of any small change in procedures can be quickly calculated.

An example of how use of this concept can prevent unnecessary regulation occurred some 8 years ago. A very careful PRA was done by the Nuclear Regulatory Commission (NRC, 1990) for a number of "typical" nuclear power plants including an early Boiling Water Reactor (BWR). In all cases it was found that the core melt probability was LESS than one in ten thousand per year. There are uncertainties about this calculation, and there has been some discussion about whether one should take the median, the mean or the mode of the probability distribution. I have argued elsewhere that one should take the mean, and do so in what follows. An immediate use of this argument was discussed at a NRC research advisory group meeting. The NRC staff was suggesting addition of safety devices to BWR Mark I reactors to improve safety. I, as a member of that advisory committee, pointed out that these reactors met the safety goal with flying colors. Either the safety goals were wrong, or the NRCs research program that produced NRC 1150 was useless, or the staff suggestion was excessive. The committee agreed with me and so did the director of regulation. In this case the staff suggestion was dropped. Unfortunately this was an isolated instance. It was also an instance in which regulation was not increased rather than an instance in which regulation was actually decreased. Reducing regulatory requirements is FAR more difficult. However, I urge that NRC have a formal and MUCH more rapid procedure for examining regulations.

In 1991 I was Chairman of a task force reviewing the safety of the nuclear power plants in Taiwan on behalf of the Minister of Foreign Affairs. The director of regulation in Taiwan told us that he accepted the idea of guidelines but wanted to have the core melt frequency to be less than 1 in 100,000. I asked why he wanted it to be so low when the careful studies by NRC thought that 1 in 10,000 was low enough. The reply was that "industry can meet it". Maybe so. In the event, I believe that Taiwan did NOT change the safety goal. Since the power plants, which were US designed and very well run meet the 1 in 10,000 goal easily that leaves wiggle room for the utility company (TAIPOWER) to cope with occasional lapses of their staff and an occasional overzealous regulator. The ROC AEC can, by using PRA can easily justify their goal to the public and TAIPOWER can proudly tell the public that they are doing even better.

Hard though it is to reduce the severity of a regulation, it is harder to forgive a deliberate violation of regulations even when that violation does not result in any safety goal being exceeded. But again I urge immediate and rapid effort in this direction. If there has been a procedural violation the NRC must of course act in some way because such violations can escalate. But I suggest a graded response. The power plant might be shut down, as were the four power plants at Millstone and Connecticut two years ago, when a "whistle blower" pointed out that the technical specifications had been routinely violated. But the shutdown should only have lasted until NRC can determine whether or not the violation led to exceeding the safety goals. With fast computers a PRA can be set up to do such an analysis within a week or two at most. If a safety goal was NOT violated, it seems a clear indication that the regulation or technical specification was too strict and it could be modified and the reactor allowed to restart with no further "punishment".

The US NRC was spawned in 1974-5 from the old Atomic Energy Commission to separate the promotional role of nuclear energy from the regulatory role. It was already geographically separated by putting the promotional arm of the AEC in Germantown and the regulatory arm of AEC in Bethesda. But unlike the mandate given to the AEC by the Atomic Energy Act of 1945, the NRC has no mandate to keep power plants in operation - only to ensure that the power plants operate without undue risk to the public. It was left to ERDA and now the Department of Energy to promote nuclear energy and to provide the balance. It is important to realize that the utility companies cannot and will not by themselves perform this function of balance. The utility companies are under close local or regional control, and historically have shown extreme reluctance to challenge any regulatory body. There is a great unbalance in power. A regulator often has the ability to keep a power plant shut down for an extra day - an action which costs the utility company $1,000,000 per day. There is no counterbalance to ensure that this power is used wisely and well. The Nuclear Regulatory Commission has been sued in the courts, (in what seems to be the preferred procedure in the USA for obtaining balance) by one or another group opposed to nuclear power, but to the best of my knowledge has not been sued by utility companies. Any regulator will automatically adjust his strategy to minimize lawsuits - and probably that is easiest done by ensuring that the number of lawsuits from each side is equal. If there is no one actively promoting nuclear energy, therefore, the regulation will inevitably become more strict and will force unnecessary price rises until price competition destroys the industry.

The NRCs recent record in the above respects has been very bad. I mentioned above that the radwaste regulations cost 1000 times too much. The actions at Millstone point were vastly exaggerated also. Over 2 years ago I asked the Chairman of the Commission by FAX what the technical problems were and what effect they had on safety. I have had no reply but after 2 months I did get a reply from the Director of Regulation, who since resigned, who gave me a 2 page comment on procedural violations. NO ONE within or without the commission has challenged my contention, repeated many times since then that the effect of the procedural violation that caused the shut down was a change of at most 1 in 100,000 per year in core melt probability. The NRC should have been able to realize within 2 weeks that this is one tenth of their safety goal and that the draconian action of a two year shutdown was unnecessary.

The cost of the over regulation at Millstone is huge and seems to have been deliberately understated in many reports so far. I take it here to be the busbar cost of replacement electricity of about $3,000,000 a day or 2 billion dollars so far. The effect on public health is also huge. Supposing the replacement electricity to come from a mixture of fossil fuels and hydro power in the average proportions, each power plant replacement costs over 50 premature deaths a year from air pollution, (Wilson and Spengler, 1996) or over 400 deaths so far. Indeed it is calculations such as this that led former Senator Tsongas to declare "I do not see how anyone who prefers burning coal to nuclear power can call himself an environmentalist".

In addition to the above we must also remember the cost that regulatory delay can provide. The license hearing for the Maine Yankee construction permit was only 6 hours long. The operating license hearing a few years later was only 2 days long. More recently some hearings have lasted 10 years. Temporary storage of high level nuclear waste and uranium isotope separation are inherently safer than operating a reactor - the latter much so. Yet the hearings for storing waste in the Goshute Indian reservation are optimistically estimated to last 4 years and a plans for a uranium isotope separation facility in Florida were abandoned after 7 years of no progress. These procedures are largely under the control of NRC and could be accelerated. But there are indications that NRC may be changing. Shirley Ann Jackson (Chairperson of the NRC) stated in October 1998, "Regulation, by its nature, is a burden but ... the burden must be made clear, based on risk insights, with performance expectations clearly defined." And, Ashok Thadana, Director of Research states, "It's time to make the regulatory process more efficient." Let us hope that these fine words lead to rapid actions.

Regaining Regulatory Balance

How can we regain the balance in regulation? I submit that two steps are necessary. The first is a procedure to decide to regulate nuclear power in a more efficient way (including deciding upon how much regulation is necessary). This would NOT be active promotion of nuclear power. The second necessary step is to find a group which will play the active promotional role that is so necessary in the US system and those patterned after it. I suspect that is not now politically possible for the Department of Energy (DOE) to be such a group to that extent the political concept of 1973 when the AEC was broken up was fatally flawed. Other mechanisms must be found. Who should constantly call the regulator to task when he takes actions that exceed his own goals. In the USA that would have to involve lawsuits because that is where the action finally occurs in any subject.

In 1987 six Long Island residents started a lawsuit nominally against Long Island Lighting Company but really against New York State, (with a supporting brief by a Department of Energy more friendly to nuclear power than the present department) in an effort to prevent them dismantling the Shoreham nuclear power plant without filing an environmental impact statement - and in that statement would have inevitably had to contradict the comments of Governor Cuomo's staff in the previous EIS that nuclear power is environmentally advantageous. This suggests to me that a publicly minded group of scientists, who are concerned about the three environmental issues with which I began this paper should form a group to be watchdogs and file suit to enforce fair regulation when appropriate. They might demand that NRC modify their regulations so that they do not demand the expenditure to reduce radiation exposure of more than their own guideline of $1,000 per ManRem. They might demand that no regulation be promulgated, no old regulation be enforced and no draconian action be taken against a utility company to reduce the frequency of core melt below the NRC safety goals and criteria. Unless something is done, I do not think that nuclear power in the short term will survive.

I urge NRC to begin once again to act in the public interest and according to their own safety goals. I urge NRC to change regulations to in either direction to match these goals. When there is a procedural violation that has safety consequences that are within these goals to give a slap on the wrist rather than an execution. I urge each commissioner to take the approach of W.S Gilbert's Mikado who made it "an object all sublime.... to make the punishment fit the crime". Let us hope that it will be achieved in time - before the nuclear industry is destroyed.

The Utility Company Failures

Of course utility company staffs and in particular utility company managements are often the most to blame. We have seen in the late 1970s how TVA went in a few years from one of the best utility companies to one of the worst. In the early 1990s Ontario Hydro went from having the highest plant availability of any reactors in the world to being among the lowest. Many observers attribute each of these to a change in top management. There is less agreement on whether the management was malevolent (antinuclear) or merely incompetent. Northeast Utilities had clear management problems. The costs were one of the highest and in an effort to reduce costs there was, and is a temptation to cut corners. In addition there are cases of inadequate regulation. That usually comes from inadequate alertness. There are also occasional "whistle blowers" who for whatever reason raise issues that they feel have been neglected. The Ontario hydro board used to have one person who understood nuclear technology - now it has none. (But the management incorrectly in my view insist that the only problem is on the shop floor) But the regulatory structure should be able to cope with this. If it is necessary (and I do not believe it is) to always have perfect regulators and perfect management to run nuclear power, there is no hope that costs can be reduced. Fortunately the PRA confirms for us that light water reactors are a forgiving technology. When there is public concern, there is a temptation for a regulatory authority to tighten up all round in the hope of reassuring the public. Indeed several successive Chairmen of NRC recently seem to have felt it politically desirable to do something dramatic to tighten up regulations in an attempt to convince the public that they are no pushover for industry. I do not believe that it does reassure the public. I believe it makes matters worse, by implying that the previous regulatory action had been too lax. Far better would be to study, know, understand and explain to the public the small effect that such problems have on safety. As noted above, we now have the techniques of PRA available to ensure the completeness which is otherwise so difficult. I suggest a joint approach by utility companies - perhaps through the Institute of Nuclear Power Operations (INPO) - the NRC and academia (hopefully with funding from DOE) to study the critical interaction between regulator and licensee - always with a background of the risks of other energy technologies as discussed in Comparative Risk Assessment (CRA). Even though regulatory procedures vary between countries IAEA could also play an important role.

Finally I claim that we can get close to meeting the Kyoto commitments if we wish. But we must be active and not merely talk about it.

REFERENCES

Benedict, M., 1971, Electric Power from nuclear fission, *Technology Review* October/November.

Benkhe, W., 1997, Communication from W. Benkhe, former C.E.O. of Commonwealth Edison Co., owner and operator of Dresden II and Dresden III.

CMP, 1972, Report from Central Maine Power, majority owner of Maine Yankee. This cost does not include a (later) cost of $20 million to remove a causeway and improve tidal flow in the coolant estuary (which many experts thought was unnecessary and certainly would NOT and have been, and was not demanded of a fossil fuel plant).

EPA, 1998, Guidelines for Preparing Economic Analysis (draft). U.S. Environmental Protection Agency, July 17, 1998.

Federal Register, 1986, 51 FR 28044.

Hill, J., 1997, Communication to the author by Sir John Hill, Chairman of the UK Atomic Energy Authority at the time.

Kress, T., 1994, Report to Nuclear Regulatory Commission from the Advisory Committee on Reactor Safeguards.

NCRP, 1990, Implementation of the principle of As Low As Reasonably Achievable (ALARA) for medical and dental personnel, National Council on Radiation Protection and Measurements, Report No. 107.

NRC, 1975, Rule making RM - 30 – 2, Nuclear Regulatory Commission.

NRC, 1990, Severe Accident Risks: An Assessment for Five U.S. Nuclear Power Plants, Nuclear Regulatory Commission report NUREG 1150.

NYT, 1996, Figures reported in the New York Times.

Parkinson, C.N.,1957, Parkinson's law, page 12, Houghton Mifflin, Boston.

Shlyakhter, A., Stadie, K. and Wilson, R., Constraints hindering nuclear electricity," US Global Strategy Council, 1995.

Tengs, T., et al., 1995, Five hundred life saving interventions and their cost effectiveness, *Risk Analysis* 15:369.

Towers and Perrin, 1995, Report to Nuclear Energy Institute.

Webster, W., 1972, Letter from William Webster, President of NE Electric System to Richard Wilson.

Wilson, R., 1994, The potential for nuclear power, in: *Global Energy Strategies: Living with Restricted Greenhouse Gas Emissions*, J.C. White, ed, Plenum Press, NY, pp. 27-45.

Wilson, R., and Spengler, J.D., eds., 1996, *Particles in Our Air: Concentrations and Health Effects* Harvard University Press, Cambridge, MA.

Wilson, R., 1998, Remembering how to make cheap nuclear electricity, testimony to Subcommittee on Energy and Water Development, United States Senate Committee on Appropriations, Tuesday, May 19, 1998, Washington D.C.

Wilson, R., 1999, Overregulation and other problems of nuclear power, presented at the Global Foundation Conference, Washington, D.C., October 1998

NUCLEAR FISSION: FOR SAFE, GLOBALLY SUSTAINABLE, PROLIFERATION-RESISTANT, AND COST-EFFECTIVE ENERGY

R. A. Krakowski

Systems Engineering and Integration Group
Technology and Safety Assessment Division
Mail Stop F607
Los Alamos National Laboratory
Los Alamos, New Mexico 87545, USA

L. Bennett

International Atomic Energy Agency, Retired
6, rue Maillasson
F-92100 Boulogne, France

E. Bertel

Nuclear Energy Agency/OECD
12 bld des Iles
F-92130 Issy - Les - Moulineaux, France

ABSTRACT

To varying degrees, under varying priorities, and depending strongly on country/region, the advancement of nuclear energy must deal with four cardinal issues: *waste, proliferation, cost, and safety*. While solutions to these issues are usually offered within a techno-economic framework, the future of nuclear energy is being determined by these issues largely through a socio-cultural paradigm *vis á vis* dwindling public acceptance of this technology. After giving a general perspective of energy consumption in terms of minimal requirements, needs, and constraints, the status of and prospects for nuclear energy are reviewed. Key elements of the public acceptance issue are then described, along with pathways being suggested to improve acceptance of this technology. After outlining a range of technical approaches to the future of nuclear energy offering new directions that might improve public acceptance through addressing the four cardinal issues, the long-term and global impacts and trade-offs for nuclear energy as estimated from an energy-economics-environmental model are reported, with an emphasis being placed on how that future may be tied to present concerns over the growing use of fossil fuels and possible climatic impacts related thereto.

Preparing the Ground for Renewal of Nuclear Power, edited by
Kursunoglu *et al.* Kluwer Academic / Plenum Publishers, New York 1999.

INTRODUCTION

Attracted by an abundant, inexpensive, and clean-burning fuel, energy from nuclear fission in the 2-3 decades following World War II saw an enormous growth in commercial application; the worldwide rate of additions in nuclear-electric capacity reached ~30 GWe/yr in the mid-1980s. That growth has decreased to a trickle of a few GWe/yr, with major industrial countries either experiencing stagnation, planned phasing out, or outright moratoria for the use of nuclear power. The uranium that feeds the nuclear fuel cycle is both more abundant than first thought and remains inexpensive because of diminished demand relative to earlier expectations. The costs of building capital plant needed to convert this relatively inexpensive fuel have increased, while both fuel and capital costs of the fossil-fuel competition have diminished. Additionally, the image of pristine conversion of nuclear energy into clean electricity has been dulled by the prospects of accidental releases of dangerous materials that in (a portion of) the public's eye links civilian nuclear energy to the military dark side of things nuclear. Furthermore, civilian nuclear energy generates by-product materials of military interest which create a situation that draws further attention to that dark side. Lastly, the longevity of the hazards associated with wastes created as a by-product of nuclear fission, and the inattention given so far to the long-term implications of that waste steam, have added further force to tilt the scales by which the public assays cost *versus* benefit. Although a few energy-strapped industrialized countries continue to sustain the meager worldwide growth in nuclear energy, four cardinal issues - *waste, proliferation, cost,* and *safety* - loom large, intermixed, convolved, and unquantified in driving key public decision processes that determine from whence the next exaJoule of primary energy will derive.

The large combinations of materials and configurations available to perform essential functions in a nuclear power plant (*e.g.*, efficient stewardship of neutrons sustaining the chain reaction, neutron moderation required of thermal-spectrum configurations, coolants, structural alloys, *etc.*) offer a rich array of options to harness safely and economically nuclear energy. Although the approaches to nuclear power plants have narrowed significantly over the last three decades, these material/configurational options remain available to provide engineering resolution to the aforementioned four cardinal issues. Can commercial approaches to nuclear energy be "re-engineered" in ways that lead to safer, globally sustainable, proliferation-resistant, and cost-effective nuclear power plants? – certainly yes. Will such "re-engineered" nuclear power plants be received by increased public acceptance? - probably not, possibly yes, depending on country, region, and culture. The issues driving the demise of nuclear power generally fall outside the realm of technology and economics; the issue of (growing) public (un)acceptance has social, cultural, and political origins, albeit, a re-discovery of the technological, economic, and environmental benefits claimed earlier for nuclear energy is an essential ingredient in the process of regaining public acceptance of this technology. Against this backdrop, after giving a general perspective on the role and demand for energy and the niche the nuclear energy might fill, the status, prospects, acceptance, and possible future(s) for nuclear energy are discussed. Sample results are then given from analyses of impacts and trade-offs associated with a range of supply-side and demand-side (and combinations thereof) scenarios based on a global (thirteen-region), long-term (~100 years) energy-economics-environmental model; possible roles and limitations of nuclear energy in mitigating global warming are reported, with a detailed elaboration being given in Ref. 1.

PERSPECTIVES

Energy facilitates all human endeavors (heating, cooling, illumination, health maintenance/enhancement, food production/transport/storage/processing, education, industrial processes, transportation).[2] The four revolutions that mark major advancements in the human condition and population (*e.g.*, Discovery of Fire; Invention of Organized Agriculture; Creation of Cites; Scientific Discovery and the Ensuing Industrialization), as well as the impending Fifth Revolution (Decarbonization of Global Metabolic Cycle Driven by Resource/Environmental Limitations),[3,4] have all been accommodated in part by shifts in the composition or mix of energy usage (*e.g.*, human, draught animals, wood, water, coal, petroleum, gas, nuclear, solar, *etc.*) as well as the amount or intensity of energy used to create, maintain, or shift the human environment (*e.g.*, energy intensity or energy use *per* unit of productivity). As humankind progressed through these four revolutionary phases, linkages between population, energy use, and productivity (as measured by generally improved well-being or human condition) evolved from the very weak or non-existent to the very strong and increasingly complex, interconnected, and evolving. These growing connectivities between population, energy, and productivity, as is illustrated heuristically in Fig. 1, have enlarged the role of energy as a direct facilitator of human endeavors by incorporating catalytic elements related to economic and societal development, including population growth.

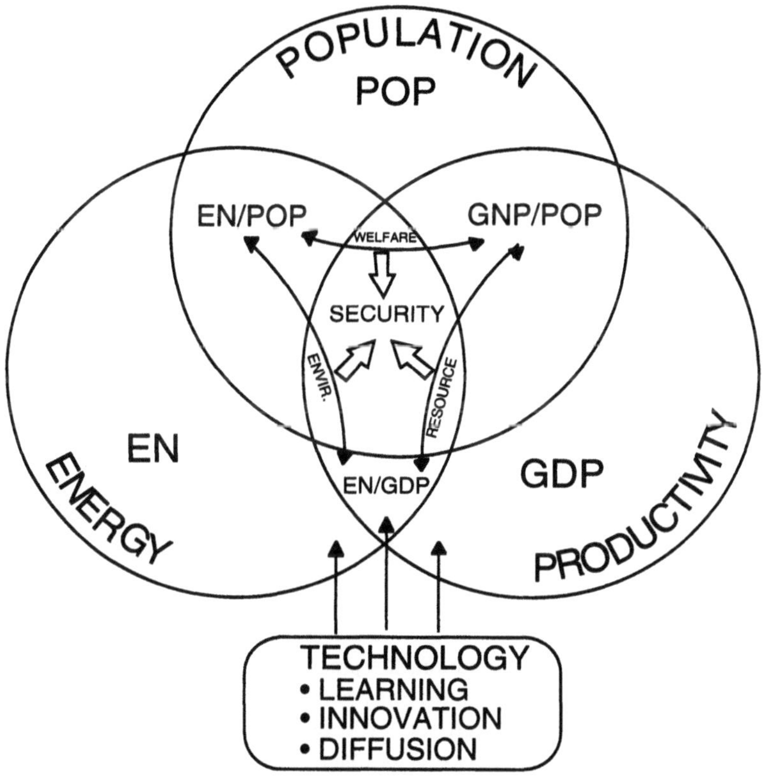

Figure 1. Relationship between energy and societal-economic-political infrastructure impacts.

Ever-growing and inter-related energy networks have evolved to fuel an ever-growing and inter-related societal-economic-political infrastructure. Although far from complete, the outline of Chapter 2 of Ref. 2, as represented in Table 1, illustrates key infrastructural impacts and the facilitating role played by energy in general. With the crowding and competition at key nodes of global energy networks that is accompanying internationalization and globalization in a world of evolving economic multipolarity, security of resource, environment, and regional well-being (Table 1) become essential actors in a kind of global stability play. The linkages illustrated in Fig. 1 tighten and the resource and environmental limitations become more evident as the Fifth Revolution proceeds, and understanding of these relationships for the purposes of setting productive and effectual policies, as well as understanding possible consequences of a range of policy alternatives, becomes crucial. This understanding can be enhanced through long-term global models that describe possible futures shaped by the kinds of interactions depicted heuristically on Table 1 and Fig. 1.

Table 1. Heuristic Listing Showing Strong Coupling Between Population, Productivity, Energy, and Security-Related Impacts on Welfare, Resources, and Environment.[2]

Interaction	Infrastructure Impact/Connection
Energy-Society	Poverty Population Food
Energy-Environment	Health Local Pollution Global Climate Change Land Degradation
Energy-Economy	Investment Requirements: • Supply/Generation • Conservation/Efficiency Foreign Exchange/Imports Price/Demand/Productivity
Energy-Security	Trade/National Security Nuclear Proliferation Cross-Boarder Pollution Technology Transfer/Diffusion

As the globe enters the Fifth Revolution, the present 5.7 billion inhabitants are expected to more than double in the next half century;[5] only half of the present population could be fed at consumption levels presently enjoyed (on average) in the U.S.;[6] and 30% of present population annually utilizes 70% of the world primary commercial energy, giving a disparity in *per-capita* energy consumption that exceeds a factor of 7 between OECD and developing (DEV) countries (25 GJ/yr/*cap versus* 180 GJ/yr/*cap*, with this factor exceeding 20 at the *extrema, e.g.*, India *versus* the United States). Fig. 2 compares[6] the *per-capita* energy consumption with population for industrialized (IND = OECD + FSU/EEU) and DEV countries, along with projected energy increases driven by both population and *per-capita* primary-energy (living-standard) increases. Short-term growth in primary-energy demand[7] lies in the range 1.9-2.4%/yr, with the short-term demand for electrical energy (presently 35% of primary energy) growing at a much faster rate (2.4-3.2%/yr); within 50 years, primary energy directed to the generation of electricity could easily comprise 40-45% of total primary energy (an increase of 50-100% from today).

A complex, regionally dependent, and interconnected array of forces will determine what fraction of the 30-40,000 TWeh/yr in 2050 will be of nuclear origin (presently 17.3% of total global generation, or 2,312 Tweh/yr in 1996)[8], as well as the role (if any) that nuclear energy will play in providing non-electric energy needs in the 2050 time frame. Table 2 summarizes key drivers of nuclear-energy demand, with Ref.-9 elaborating on these drivers as applied under constraints imposed by the goal of achieving "sustainability" relative to other energy sources. Sustainability, under these condition relates to providing reasonable assurance that the needs of future generations are not placed in jeopardy by the creation of toxic wastes and pollutants, while minimally exploiting natural resources in the course of providing contemporary services and products. In this context, the results of net-energy balances (total life-cycle energy input divided by output) and carbon-dioxide emission intensities (again, expressed on a life-cycle basis) are given in Figs. 3 and 4, respectively,[6,10] as high-level comparative indicators for a range of conventional, renewable, and nuclear energy sources. Given that the four cardinal issues for nuclear energy can in fact be addressed satisfactorily, and taking into account the significant differences in sheer mass of waste generations, sustainability indicators in this regard bode well for nuclear energy.*

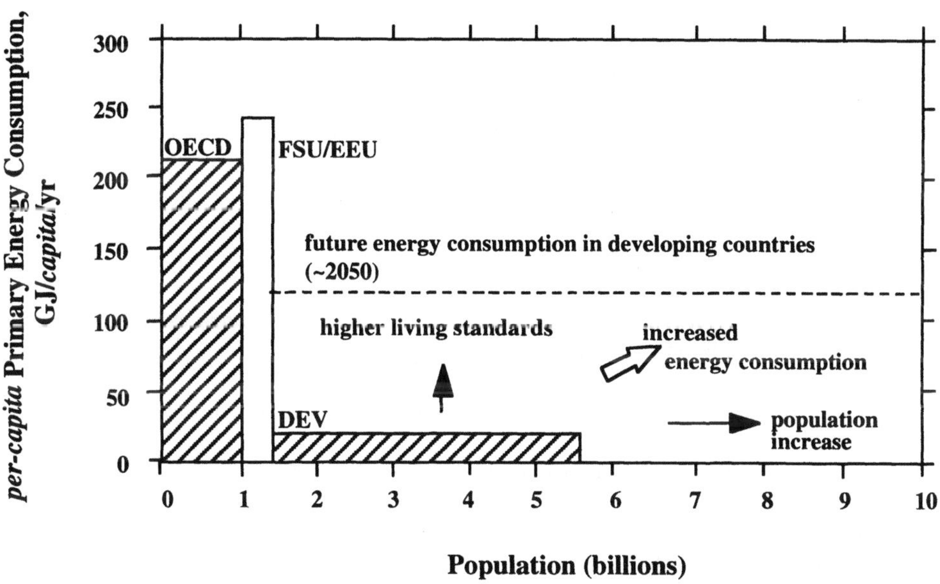

Figure 2. Population and *per-capita* primary-energy consumption,[6] showing impact of ~50-year growths in both.

---

*A 1-GWe coal plant each year generates 0.35 Mtonne of solid waste (which contains ~400 tonne of highly hazardous heavy metals), 6 Mtonne CO_2, (1.6 MtonneC), 17 ktonne SO_2, and 4 ktonne NO_x (with abatement systems). These waste masses compare to a 1-GWe once-through (no recovery of used fuels) nuclear plant that generates 400 tonne of low-level waste, 200 tonne of intermediate-level waste, and 20 tonne of high-level waste.

Both the level of *per-capita* energy consumption and the efficiency with which primary energy is converted to provide desired or required energy services (final energy) are strong determinants of increased energy consumption depicted (for 2050) in Fig. 2. The general increase in *per-capita* energy consumption tracks the historical reduction in (exhausting, often-dangerous, sometimes child/youth-intensive) human labor that has driven the agricultural → manufacturing → service-oriented cultural transformations, and the economic transformation related thereto. While the average *per-capita* (primary) energy consumption today amounts to ~50 GJ/yr/*capita*, inequalities among nations are enormous (10-15 GJ/yr/*capita* for India, 300-330 GJ/yr for the North America). Clearly, bringing the globe to a *per-capita* level of primary-energy consumption equal to that of North America would require an amount of total annual consumption equal to the full frame of Fig. 2. A question arises as to what level of *per-capita* consumption is required for an "acceptable" quality of life. This question of socio-cultural efficiency is more central to projecting future energy needs than is improved technological efficiency *per se*. While no single meaningful indicator can measure the multi-dimensional relationship between energy use and quality of life, (*e.g.*, personal well-being, condition of environmental and social settings, intellectual aspects of human development, subjective *versus* objective valuations), little uncertainty exists with

Table 2. Drivers of Future Demands for Nuclear Energy[9]

- National and international economic contexts:
  - deregulation and unleashing of free-market forces;
  - privatization and drive for increased efficiency;
  - shift of payees from taxpayer to stockholders;
  - shrinking time frame for expected payments/profits;
  - capital (long-term) *versus* O&M/fuel (short-term) charges;
  - preferences for up-front (long-term capital) costs *versus* insensitivity to fuel price changes (short-term);
  - prospects to export (nuclear energy) equipment and services.

- Competition of alternative energy sources;
  - willingness to import energy (fossil, nuclear, electricity *via* grid)
  - levels of R&D expenditure (reduced under short-term goals of privatization, reduced government role/revenues, drive for increased efficiency)

- Role of localized (small) plants *versus* long-distance transmission and large base-loaded units.

- High value added for domestic generation in countries without access to fossil or electric-grid energy;

- Level of domestic energy sources:
  - importance of energy independence;
  - cost of fossil resources and fear/concern of decrease/exhaustion;

- State of infrastructure required for large, complex, hazardous technologies:
  - degree to which existing infrastructure is atrophying, decaying;
  - need/potential to build or import required infrastructure;
  - magnitude of mismatch between need and availability of infrastructure;
  - constraint of energy choice and development, role of IAEA.

- policy and/or social/cultural forces driving degree to which (energy) externalities are included in the costs associated with a given energy technology.

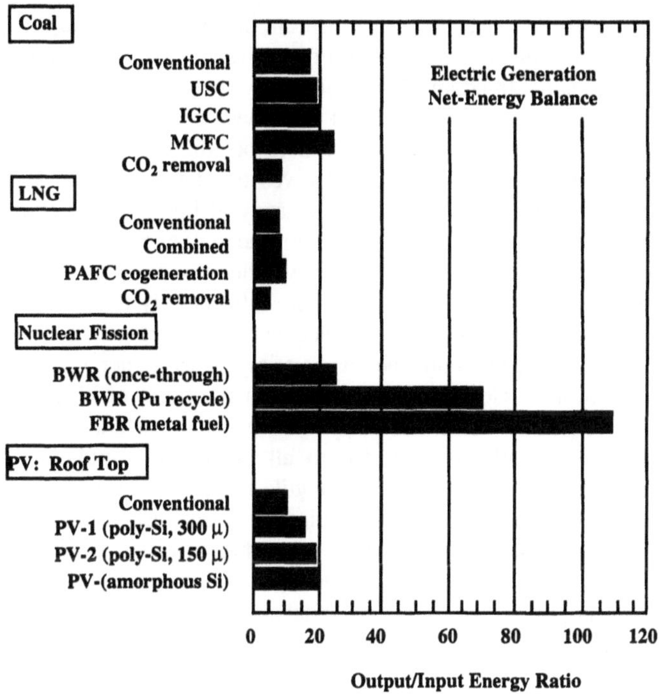

Figure 3. Results of net-energy balances (total life-cycle energy output divided by input) for a range of conventional, renewable, and nuclear energy sources.[6,10]

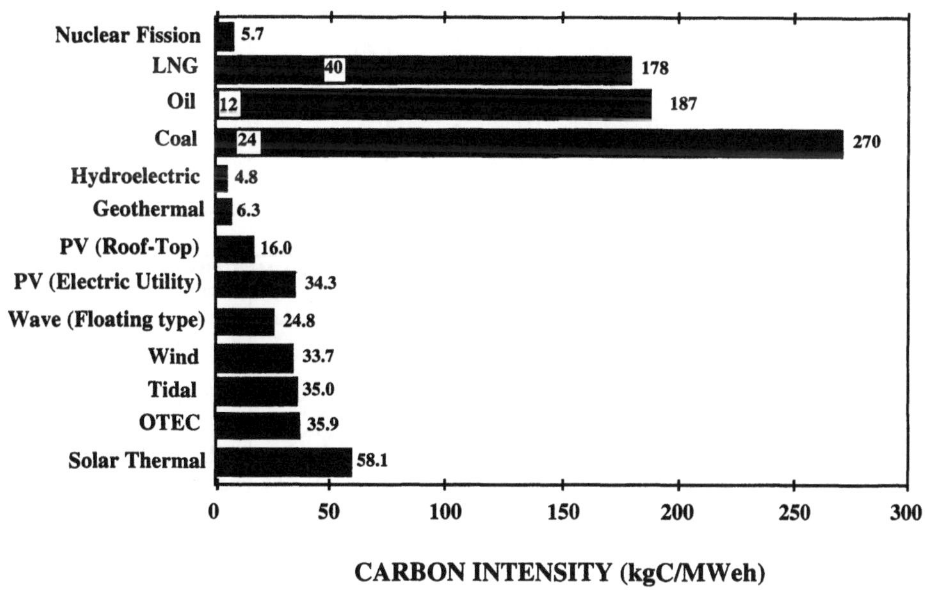

Figure 4. Results of carbon-dioxide emission intensities (expressed on a life-cycle basis) for a range of conventional, renewable, and nuclear energy sources.[6,10]

respect to the relationship between the level of *per-capita* primary-energy consumption and the physical quality of life (*e.g.*, health care, nutrition, housing, education).[11]

Furthermore, the *per-capita* energy costs of providing intangible components of "the good life", including personal freedom (democracies have been incubated under conditions of very low *per-capita* energy consumption; centrally planned economies have tended to operate with high *per-capita* energy consumption[11]), tend to be low compared to that required to provide high physical quality of life. Figure 5 gives the dependence of five indicators of physical quality of life (infant mortality, life expectancy at birth, total fertility rate, literacy rate, and food availability) on *per-capita* primary (commercial) energy consumption. The first two indicators are the most revealing,[11] in that they reflect the impacts of long-term health, nutritional, and environmental care. The literacy rate reflects only the percentage of children that complete primary school, and does not account for functional illiteracy. The food-availability metric reflects mainly abundance and variety, with higher values not necessarily reflecting "better" conditions. Generally, all metrics reach a point of diminishing returns in the range 40-50 GJ/yr/*capita*, which fortuitously is close to the world average. While a "good (physical) life", therefore, could be provided to all the world populations by equalizing present energy consumption, deep cuts in *per-capita* energy consumption by the wealthy nations would be required. Furthermore, this point of diminishing return may be pushed to higher values of *per-capita* energy consumption when higher-order measures of life quality are included [*e.g.*, tertiary education, dealing with (economic) productivity shifts related to and caring for aging populations, *etc.*]. Nevertheless, setting long-term, global *per-capita* energy consumption goals a factor of 3-4 below the upper extreme of the present distribution, while achieving acceptable life-quality levels (*e.g.*, increasing social/cultural life-quality efficiency) will go much farther than any increase in technological efficiencies.

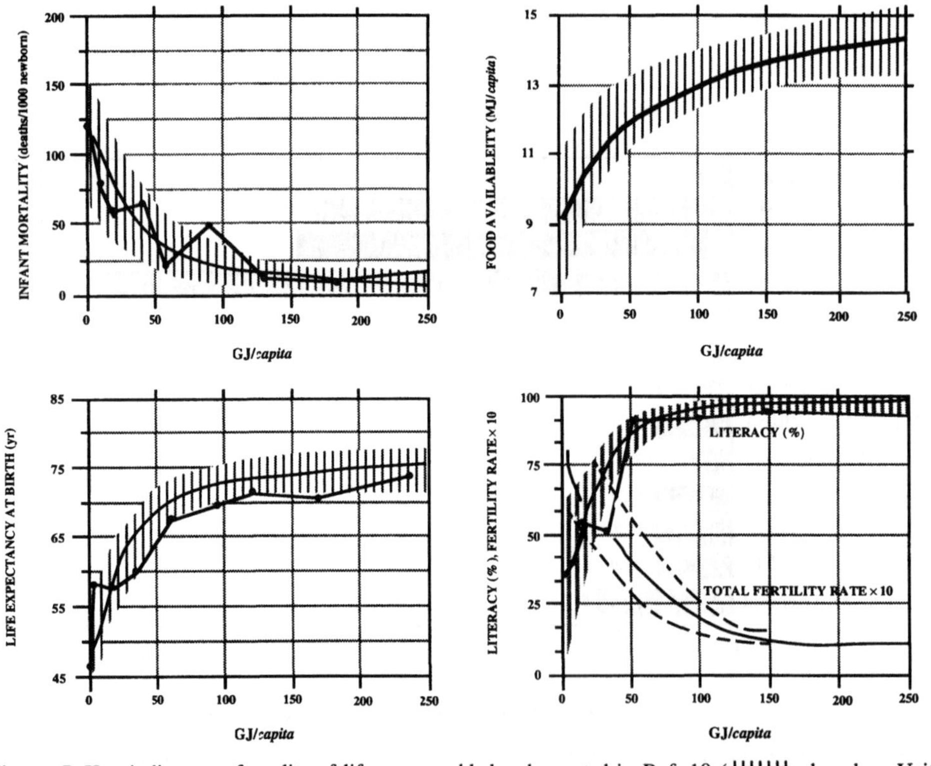

Figure 5. Key indicators of quality of life as assembled and reported in Ref. 10 (┼┼┼┼┼┼, based on United Nations[12] and World Bank[13] databases) and Ref. 14 (—●—, based on World Energy Council[15] and United Nations[12] databases).

STATUS

Introduced over four decades ago into the commercial electricity market, nuclear energy presently provides 17.3% of the global electrical energy supply;[8] on a basis of total primary energy (PE), nuclear energy (NE) contributes ~8%. A total capacity of 351 GWe is represented by 442 nuclear power plants (NPPs) operated in 32 countries. The mix of these reactors (in 1993) gives PWRs (55%), BWRs (21%, PWR/BWR = 2.7), GCRs(9%), PHWRs (7%), water-cooled/graphite-moderated reactors (4.5%, all in the FSU), and other kinds of reactors (3.5%). Nuclear energy provides over 40% of the electricity to 9 countries and 20-40% of the electricity supply in 10 countries. The 2,312 TWeh of electricity generated by NPPs in 1996 avoided approximately 8% of global carbon emissions. Nuclear energy was introduced rapidly in the 1960s and 1970s. No new orders for NPPs have been placed in the United States since 1978, leading to that capacity being pegged at 98.8 GWe (highest of any country, 28% of world capacity). Figure 6 indicates the impact of a 40-year plant-life license expiration schedule for key NE countries. Under conditions of capacity replacement coupled with a medium NE growth rate,[17] Fig. 7 illustrates the level of new NE capacity addition under conditions of moderate growth.

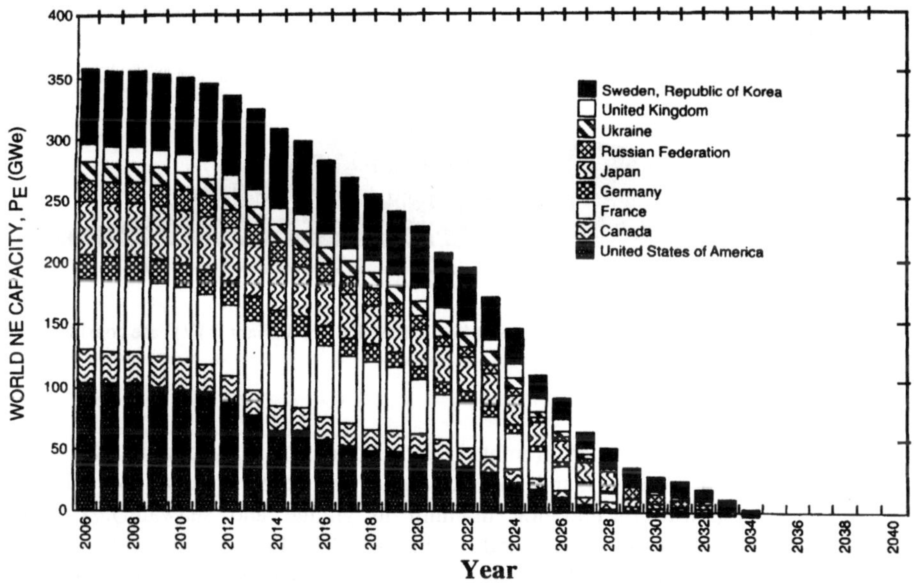

Figure 6. Expiration schedule for existing NPPs with capacity in excess of 500 MWe and for a 40-year plant lifetime.[16]

While variable from region to region, the overall operational (and economic) performances of NPPs have improved over the last few decades,[18] as measured in terms of: a) collective radiation dose *per* unit of generated electrical energy; b) the dramatic decrease in radioactive liquid releases (without tritium) *per* unit of generated; and c) the number of accidents with significant radiological consequences *per* unit of generated energy. The economic performance of NE also varies from country to country. A recent survey[19] of 15 OECD countries (Belgium, Canada, Denmark, Finland, France, Hungary, Italy, Japan, the Republic of Korea, the Netherlands, Portugal, Spain, Turkey, the United Kingdom, and the

United States) and 5 non-OECD countries (Brazil, China, India, Romania, and Russia) on the cost of electricity generation from nuclear, coal, gas, biomass, solar (photovoltaic), and wind indicates that NE on average is competitive at a (bussbar) cost of electricity of 0.032 \$/kWeh. Similar results have been reported in the U.S. for NE O&M and production cost trends,[20] which show both decreasing over the last decade, with production costs (*e.g.*, not including annual charges related to capital costs and interest during construction) being competitive.

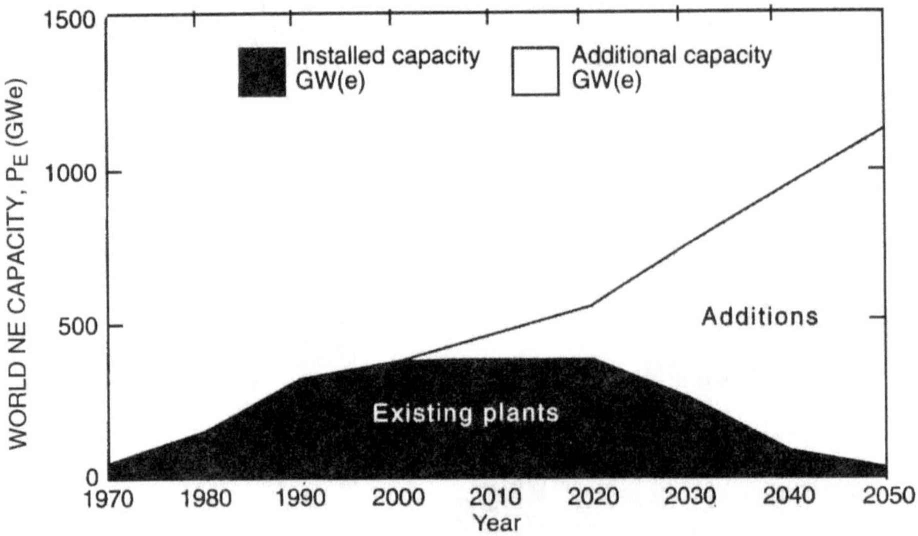

Figure 7. New NPP capacity additions required[16] according to the Medium-Variant (MV) case considered in the Ref.-17 study.

The world Proven Reserves of uranium[21] correspond to 1.5 MtonneU, Reasonably Assured Reserves are 4.0 MtonneU, and Total Resources and Reserves equal 18.5 MtonneU (excluding recoverable low-concentration uranium in granite and seawater). Complete fissioning of all ^{235}U in the total reserves and resources (18.4 Mtonne) represents an energy resource of ~10,766 EJ (1.5×10^6 EJ if all uranium were fissioned). Including the less-known thorium (^{233}U) resource would increase the 1.5×10^6 EJ by a factor of 2-3. Compared to the fossil-energy resource, the total uranium and fossil (including oil shale and clatherated methane) resources are comparable (*e.g.*, 1.5×10^6 *versus* 1.0×10^6 EJ for total uranium *versus* total fossil), but the ratio of uranium to fossil without oil shale and clatherates (97,892 EJ) amounts to ~15.3. The present world demands for total primary energy and for nuclear energy (350 GWe, ~70% plant availability, 35% thermal-to-electric conversion efficiency) amounts to 360 EJ/yr and 22 EJ/yr, respectively. At the present NE demand rate, most (90.7%) of the ^{235}U in the Reasonably Assured Reserve category (4.0 Mtonne) would be consumed in a once-through (*e.g.*, no plutonium or uranium recycle) fuel cycle by the year 2100. A linear increase in NE capacity to 2,000 GWe would increase this 100-year uranium requirement by a factor of 2.7 (9.8 Mtonne), which amounts to approximately half of the ^{235}U in the Total Reserves and Resource category.

PROSPECTS

The main growth and prospects for growth in NE is occurring in East and South Asia, with Western industrialized countries experiencing a period of stagnation in NE growth. A number of recent studies address a range of long-term NE futures. A quantitative picture of NE having to compete in a changing electric-supply-industry market is given in Ref. 22, wherein increased cost transparency, increased (short-term) market discipline, and the reduction of policy *per se* to a vestigial role is creating an environment that is very different from that into which NE entered over two decades ago. Using a more quantitative approach than that used in Ref. 22, the three NE scenarios depicted in Fig. 8 derive from the Ref. 17 study. These scenarios, in turn, were adopted from the 1995 IIASA/WEC study[23] to suggest three possible NE growth scenarios. The High Variant (HV) assumes a high overall growth in GDP and energy consumption, no limits to human technological ingenuity, no environmental (*e.g.*, GHG emissions) constraints, and large-scale use of renewable energy (RE) and NE resources; by the year 2100, the HV scenario has almost equal reliance on NE, natural gas, biomass, and renewable energy (mainly solar, with some wind and "new" RE). The Medium-Variant (MV) case on Fig. 8 is an ecologically driven scenario[23] based on less (than HV) ambitious economic growth and the use of both NE and RE to achieve sustainable growth in all regions of the world by the year 2100 (including the developing countries). The Low-Variant (LV) case is similar to the MV case, but with the introduction of a politically driven phase-out of NE. The phase-out LV case is less aggressive in reducing the role of NE than that suggested by the heuristic approach used in the Ref.-22 study. None of the cases depicted in Fig. 8 requires the introduction of breeder reactors prior to 2050, and only the HV case suggests such a resource-driven need near the end of the year 2100.

A recent NEA/OECD study[24] suggests the three alternative NE development paths depicted in Fig. 9: Variant I assumes continued NE growth leading to a 1,120-GWe capacity by 2050; Variant II is a policy-driven phase-out scenario, wherein power generation from NE would completely cease by 2045; and Variant III suggests an initial period of stagnation and possible reduction (driven by early retirement of NPPs) that is followed by a revival of NE in ~2020, with this revival also leading to an NE capacity of 1,120 GWe by the year 2050. In terms of construction rate, financing, citing and land requirements, and uranium resources, each of the three variants considered by the NEA/OECD study would present feasible challenges to the NE industrial sector.[24] Improved economic competitiveness and increasing public acceptance represent the main challenges for Variant I. The maintenance of infrastructural effectiveness during the lengthy NPP decommissionings and final waste disposal that are required of Variant II present key challenges, while non-nuclear energy sources must deal with increased problems of energy and environmental security. Table 3 elaborates on concerns associated with any NE phase-out scenario. The main challenges for Variant III are related to achieving the 75-GWe/yr construction rate required after the year 2035 (compared to 20 GWe/yr for Variant I), following a two-decade period of stagnation in the NE sector.

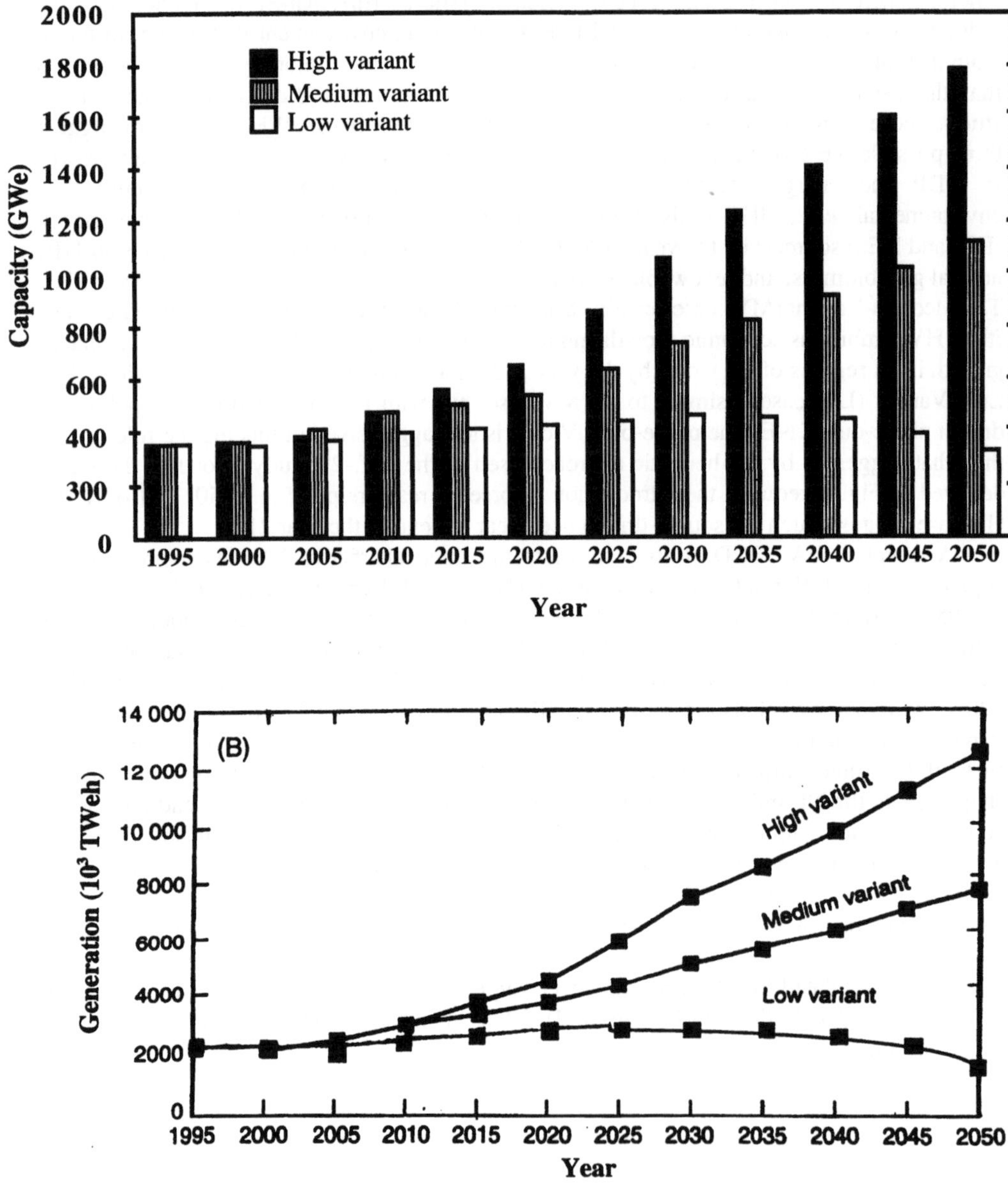

Figure 8. World NPP capacity (A) and generation (B) for three scenario variants.[17]

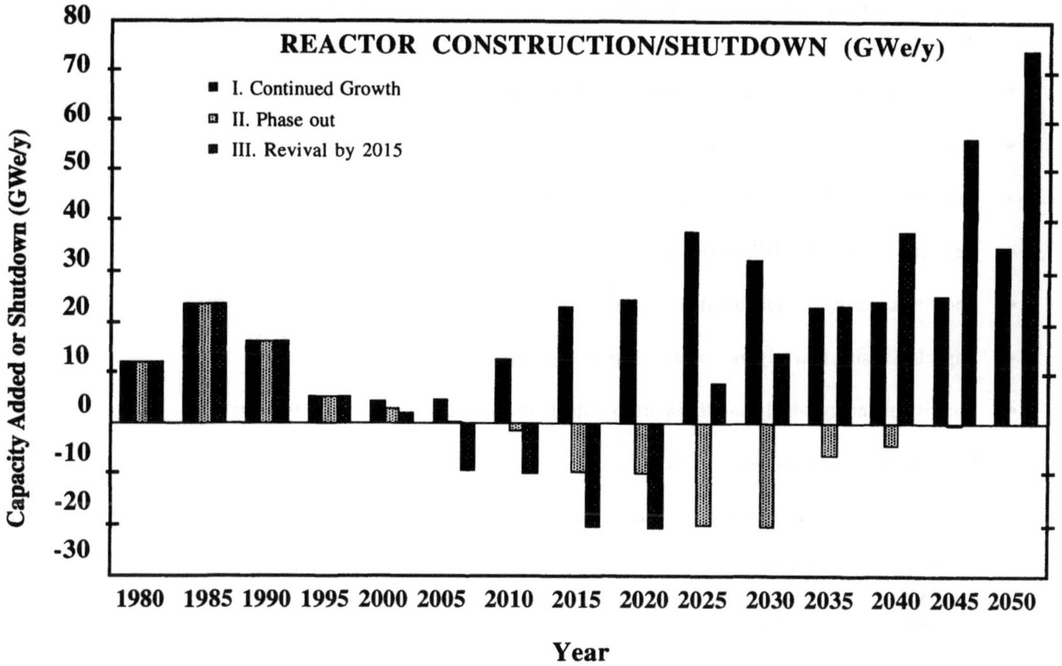

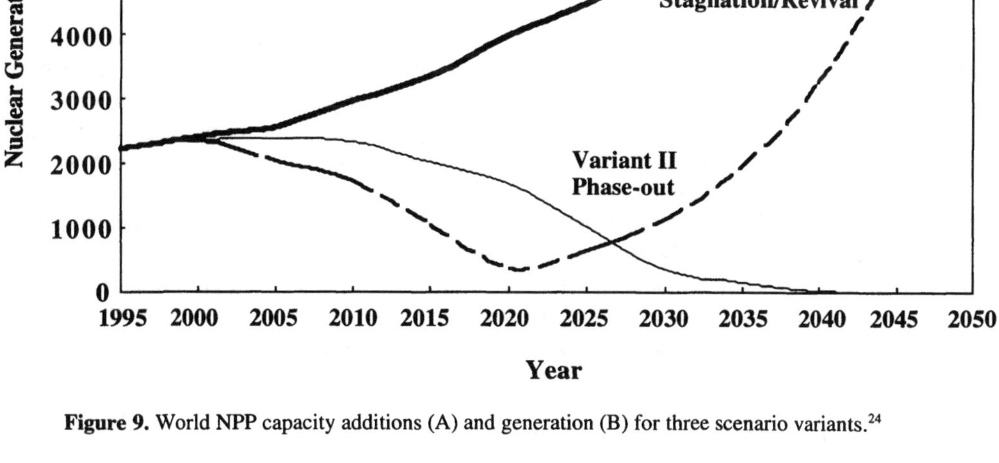

Figure 9. World NPP capacity additions (A) and generation (B) for three scenario variants.[24]

Table 3. Key Regional and Long-Term Impacts of NE-Phaseout Scenario

- Impact on NE infrastructure (*e.g.*, education, operations of facilities, regulations, enforcement, finance, *etc.*);
- Driving upward of fossil-energy (FE) prices, impact on dependent/developing countries;
- Local (cross boundary) pollution;
- Greenhouse gas (GHG) emissions and global climate change;
- Regional and global GDP impacts of the use of more expensive fuels;
- Reduced diversity of global/regional energy supplies;
- Impacts of using and relying on less-stable energy supplies;
- Security of energy supply and transport (political, accident risks, balance of payments);
- Shifts in investment strategies and costs;
- Shift in proliferation risk for legacy materials;
- General issues of stability:
  - financial (stranded capital);
  - environmental (GHG emissions, pollution);
  - energy (sources, diversity, reliances *versus* alliances);
  - military.

NUCLEAR LEGACY AND PUBLIC ACCEPTANCE

The path taken by NE, as described in the limiting scenarios depicted in Fig. 8 and 9, will be determined largely by the way in which the nuclear legacy is resolved for (by) the general public. In many countries, techno-economic solutions and "re-engineering" approaches to NEs four cardinal issues are insufficient to impact the single element that will determine the fate of nuclear power - public concerns and fears that have diminished public acceptance of this technology. The public acceptance issue is defined under a socio-cultural paradigm rather than in terms of a technological-economic one in which the expert operates. The introduction of NE as a limitless and nearly cost-free source of labor-liberating energy simultaneously with a war-making tool having nearly unlimited destructive power created a kind of public schizophrenia that combined total acceptance of the new source of energy and complete fear of its military dark side. This dichotomy of public acceptance and fear, that previously was kept at abeyance and separated, dissolved as: a) the economic benefits portended by NE advocates diminished with increased development and commercialization; b) safety concerns became better quantified and ultimately were realized; c) the fear of nuclear holocaust grew with the increasing magnitude and divergence of nuclear arsenals; and d) the level of public trust and credibility in governing and regulating institutions plummeted. The extent to which these events and concerns have affected the deployment of NPPs varies greatly from country to country and culture to culture.[25]

Nuclear energy in its relative commercial infancy has developed into a large technical system that has suffered reduced public acceptance. Without public acceptance, this technology cannot grow, even as technical solutions to the four cardinal issues are advanced.

While no single source of the public concern with NE can be identified, key societal vectors have been attached to this public concern:[25] a) the nature of risk perception by the public; b) the aforementioned legacy of fear; c) the perception of benefit; d) value conflicts and shifting cultural settings; and e) diminished institutional credibility and trust. The general public judges risk and hazard associated with a given technology differently than do experts who are well-versed in that technology. The migration of public opinion over the earlier years (1975-86) of commercial NE in the U.S. has shown[25] a strong shift from the favorable to the opposed. Results from a more recent (1995-98) survey of (U.S) public attitudes towards nuclear power,[26] however, indicate not only a significant shift towards a more positive disposition, but also that those expressing favorable opinions mistakenly thought that the majority of the U.S. public held negative views. This situation is complex, evolving, and subject to easy polarization.

The future of NE ultimately will be dictated by the level of public acceptance, and the evidence generally points to a public response to the technologies defining NE that is "value-ladened and cultural in context".[25] This condition has far-reaching implications for any effort to win increased societal acceptance of NE. In the context of a socio-cultural paradigm, multiple pathways to increased public acceptance of NE have been identified and are listed in Table 4. This approach to dealing with the all-determining public-acceptance issue revolves around adhering to a dedicated plan for breaking with the past to the extent needed to open new pathways to the resolution of socio-cultural barriers that impede technological-economic advancements.

Table 4. Pathways to Increased Public Acceptance of Nuclear Energy[25]

- Demonstrate a record of safe operation of present NPPs;
- Contain the potential for catastrophic risk:
  - continue to improve present NPPs;
  - develop new, reduced-risk and standardized NPPs;
- Separate NE from nuclear weapons (NWs);
- Re-discover the benefits of NE:
  - reduce impacts of future oil price shocks; increase energy security;
  - mitigation of GHG emissions;
  - improve competitive prices;
- Steady progress on waste management, leading to sustainable NE:
  - begin with specific waste facilities (repositories, IMRSS);
  - close the fuel cycle;
  - plan, develop, implement no-actinide, minimum-(long-lived)- fission-product systems;
- Create and implement fair, open, equitable institutions for the administration of NE.

NUCLEAR ENERGY FUTURES

Where needed, a rich array of technical solutions to the four cardinal issues facing NE can be identified. This richness of technical solution and innovation is reflected in the large combinatorial of materials available to perform the basic functions needed to generate energy from nuclear fission:[27] fuels sources (uranium, thorium), fuel types (^{233}U, ^{235}U, $^{239,241}Pu$), fuel forms (metal alloys, oxides, carbides, nitrides, fluids), neutronically compatible (in terms of neutron economy, material longevity, and waste generation) structural materials, coolants (waters, liquid metals, gases), and neutron moderators (if needed, waters, graphite). Furthermore, nuclear reactors can be "tuned" to create materials, destroy materials, or provide process or space heat, simultaneously with the generation of electrical power. The material and intellectual resources needed to "re-engineer" fission reactors for safe, globally sustainable, proliferation-resistant, and cost-effective nuclear power are not in short supply. Figure 10 suggests a nominally generic three-stage growth scenario for NE based on implementing some of these resources:[27]

- Phase I: secure existing NPPs (mainly LWRs) through license renewals (pressure-vessel life, small-component replacement); increased O&M effectiveness and associated cost reductions (NDT/robotics/remote-monitoring, coolant chemistry control, waste and dose reduction, validated reliability); reduce challenges to safety systems (optimized balance between passive control and operator intervention in matters of plant safety); (begin) gaining control of the waste issue [initiate a system of International Interim Monitored Retrievable Surface Storage (IIMRSS) systems for used fuels in preparation for Phase-III activities]; and begin reduction of separated plutonium inventories to the maximum extent practicable.
- Phase II: bridge to the future though the continued development and deployment of evolutionary LWRs [economically competitive, safer, standardized, flexible capacities, simplified, (fewer valves, fewer pumps, reduce piping, less HVAC ducting, reduced seismic building volume, less control cable) *etc.*]; meet key life-cycle requirements (close the fuel cycle under conditions required prior to attaining sustainability (*e.g.*, fissile fuel breeding); optimize balance between passive and active safety systems; address diseconomies-of-scale issues on a *per*-region/application basis [~600 MWe and expandable, grid matching, size *versus* configuration *versus* coolability, fission-product quantity *versus* number of sites, reduce (installed) capital costs, modularity (factory *versus* site fabrication, site capacity)];
- Phase III: enter into technologies required for a competitively sustainable NE future that includes proliferation-resistant breeding of fissile fuels from the world uranium and thorium resources; non-electric applications (if competitive); and either direct or support facilities that internalize and eliminate all actinides and long-lived fission products (LLFPs); in short, attain sustainability of operations from both front-end (fuel resources) and back-end (waste, proliferation risk) perspectives while assuring safety and economic competitiveness.

Specific attributes and elements of each of these three Phases are elaborated in Ref. 1, which focuses on a number of approaches to Phase III that emphasize both actinide and LLFP control; this emphasis is essential to dealing with two of the more crucial of the four cardinal issues - waste and proliferation. It remains for the technologist to assure that resolutions of these two key issues are presented while assuring good progress on the remaining two (cost and safety).

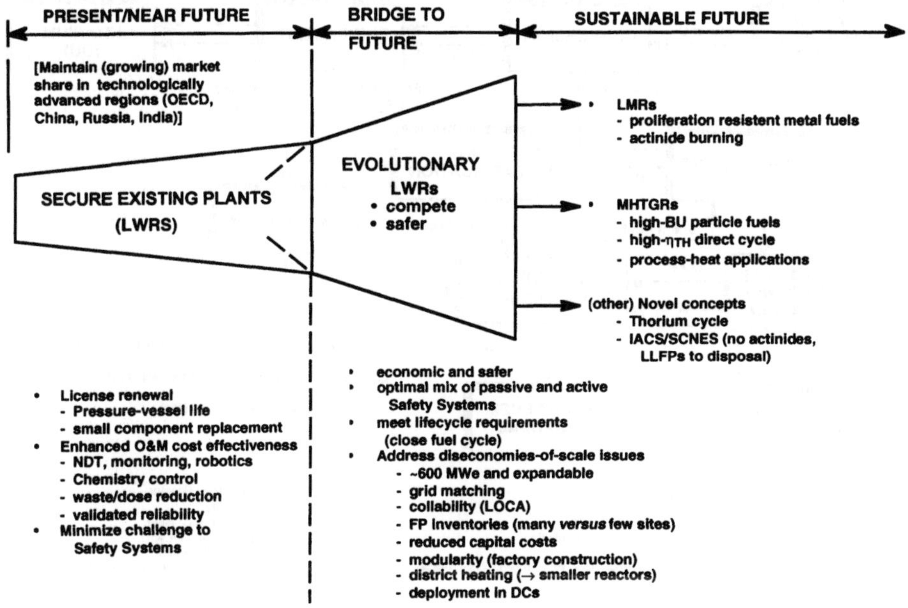

Figure 10. Possible (almost generic) growth scenario for nuclear energy.[27]

The centrum of the four cardinal issues for NE resides largely with the nuclear fuel cycle, which is illustrated in abbreviated form in Fig. 11. This diagram depicts the once-through (OT) LWR case in addition to including two recycle schemes: a) the "Closed Cycle" recycles plutonium; and b) the SCNES (Self-Consistent Nuclear Energy System) approach that recycles all actinides (ACT) and LLFPs.[28] The "Closed Cycle" based only on (thermal-neutron) LWRs are capable only of 2-3 plutonium recycles (at most) as mixed plutonium-uranium oxide fuel (MOX) before the reactivity decreases caused by shifts in plutonium isotopic mix make this fuel inefficient for use in a thermal-neutron spectrum. The harder neutron spectrum developed in an LMR (or accelerator-driven spallation neutron source)[29] is needed to fission most of the higher plutonium isotopes. Nevertheless, some decrease in plutonium inventories is possible if the MOX/LWR fuel cycle is used. For the "Continued Growth" NE scenario considered in the recent NEA study[24] (Variant-I, Fig. 9), the decrease in "non-reprocessed spent fuel" for the case where MOX fuels occupy a core volume fraction of 30% for all LWRs is shown in Fig. 12; this figure also gives the cumulative demand for natural uranium for the Variant-I OT/LWR case.

The MOX/LWR-based "Closed Cycle", while through multiple recycles degrades the nuclear-weapon (NW) attractiveness of the reactor-grade (RG, newly generated in LWRs) plutonium, it far from eliminates this material from such considerations. Furthermore, the neutron economy in a thermal-spectrum reactor (where parasitic neutron absorption rates are high) is too precarious to deal efficiently with LLFPs. The use of a fast-spectrum reactor of the Advanced Liquid Metal Reactor (ALMR) type can fission all actinides, and possibly transmute key LLFPs to shorter-lived species. Both actinide fissioning and LLFP transmutation can reduce repository (size and perpetuity) requirements, while contributing electrical power to the grid. Furthermore, the use of an Integral Fast Reactor (IFR) pyrochemical processing does not necessarily create proliferation-prone separated plutonium

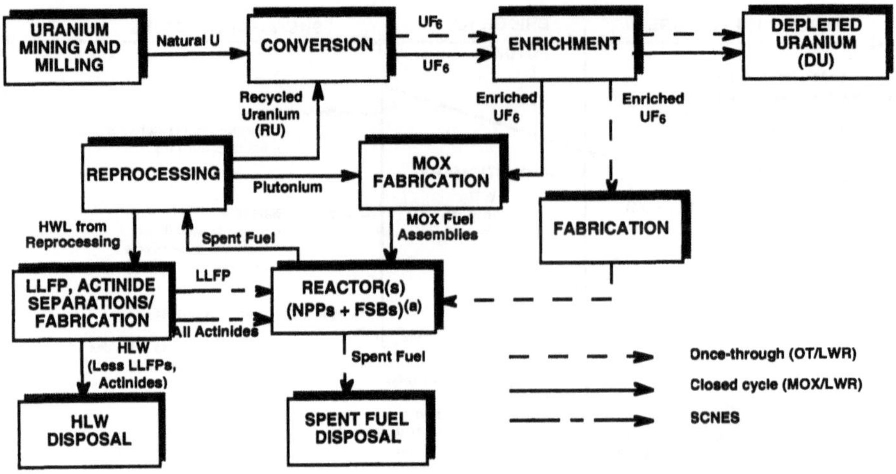

ADS = Accelerator-Driven System
DU = Depleted Uranium
FSB = Fast-Spectrum Burner (IFR/LMR or ADS)
HLW = High-Level Waste (fission products, actinides)
IFR = Integral Fast Reactor
LLFP = Long-LivedFission Products (^{79}Se, ^{93}Zr, ^{99}Tc, ^{107}Pd, ^{126}Sn, ^{135}Cs)
LMR = Liquid-Metal (cooled) Reactor
LWR = Light-Water Reactor
MOX = (U, Pu) O_2
NPP = Nuclear Power Plant
OT = Once-Through (LWR)
RU = Recycled Uranium
SCNES = Self-Consistant Nuclear Energy System

Figure 11. Once-through (OT/LWR) fuel cycle, plutonium recycle in LWRs (MOX/LWR), and an advanced actinide/LLFP burning fast-spectrum reactor (Self-Consistent Nuclear-Energy System, SCNES[28]) fuel cycle.

stream, while promising significant reductions in waste streams relative to standard (PUREX) aqueous processing methods. In pyrochemical reprocessing, chopped LWR (or ALMR metal-alloy) fuel is electrochemically transferred in an electro-refiner to a molten-salt solution, with uranium being collected in a solid cadmium cathode, and actinides (including plutonium) are recovered in a cadmium anode. The "dry" pyrochemical process has a wider potential application in other "re-engineered" configurations, and is elaborated in Refs. 1 and 29.

The SCNES approach[28] incorporated into on Fig. 11 targets the fissioning of all actinides (ACT) along with the transmutation of the seven main LLFPs (^{93}Zr, ^{107}Pd, ^{135}Cs, ^{79}Se, ^{99}Sn, and ^{129}I) to shorter-lived species. The overarching goal of the SCNES philosophy is to achieve an equilibrium (*e.g.*, non-growing for a given NE generation capacity) repository-specific inventory. Since only a single (fast) reactor is considered,[28] and neutron balance remains a concern for three of the LLFPs because of the concentrations of accompanying stable isotopes (Zr, Cs, and Pd), isotopic separations prior to transmutation are being considered. Each SCNES plant is envisaged to be fully integral, in that only natural uranium would enter the plant, and the toxicity of the material discharged from the plant to the repository would not exceed that of the feed (natural-uranium) material. The severe specifications on decontamination factors (for both ACT and LLFPs),[28] as well as criticality considerations (even for a fast-spectrum reactor) and the potential costs of isotopic separations, has broadened the scope of the SCNES concept to include accelerator-based neutron sources, and (eventually) even fusion-based systems. While the long-term goals of the SCNES system(s) are challenging on physics, engineering, and economic fronts, the attractiveness of an NE waste stream that differs little in hazard and long-term impact from that of the relatively benign input streams (*e.g.*, natural-uranium) needed to sustain the NE option can contribute to the resolution of NEs four cardinal issues in a way that promises to enhance public acceptance.

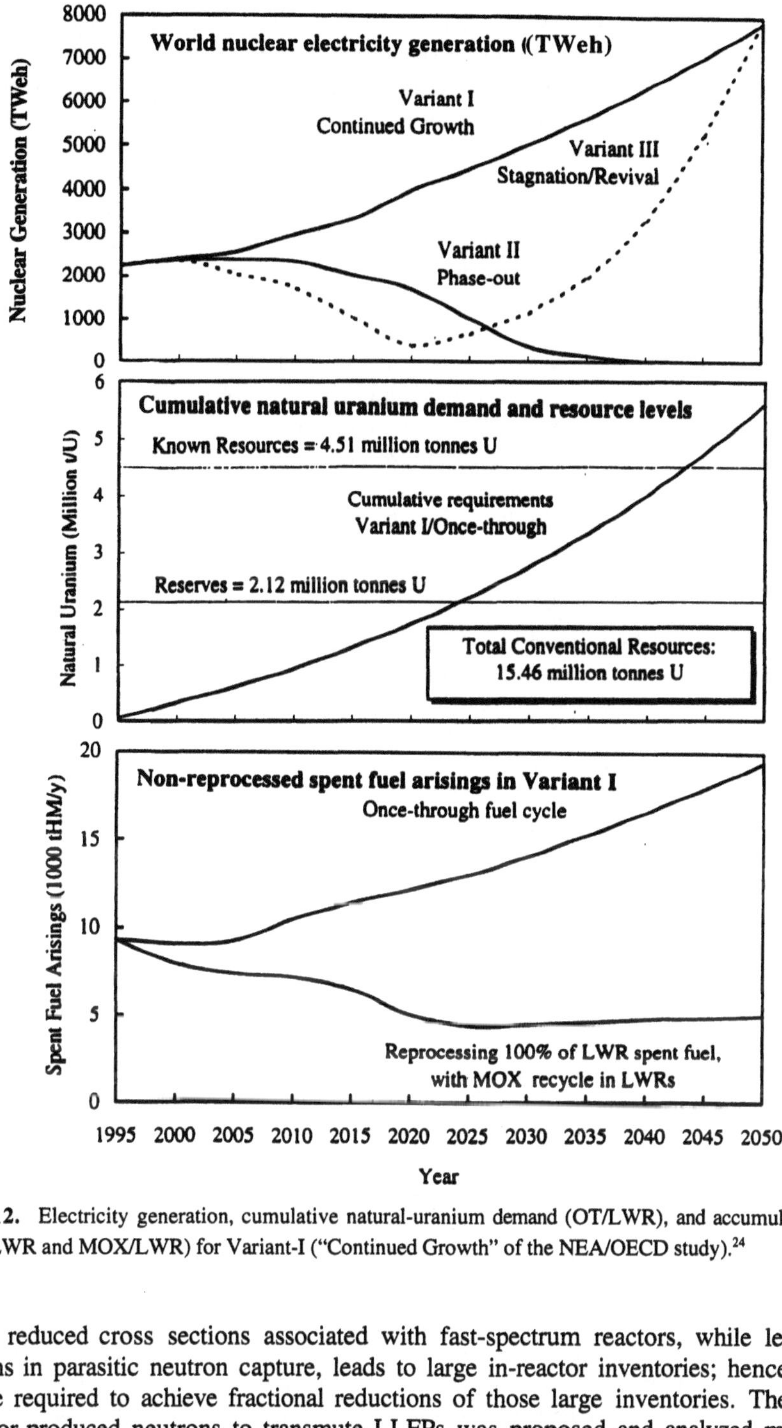

Figure 12. Electricity generation, cumulative natural-uranium demand (OT/LWR), and accumulated spent fuel (OT/LWR and MOX/LWR) for Variant-I ("Continued Growth" of the NEA/OECD study).[24]

The reduced cross sections associated with fast-spectrum reactors, while leading to reductions in parasitic neutron capture, leads to large in-reactor inventories; hence, longer times are required to achieve fractional reductions of those large inventories. The use of accelerator-produced neutrons to transmute LLFPs was proposed and analyzed more than four decades ago.[30] With advances in the prospects of high-current accelerators in recent years,[31] interest in Accelerator Transmutation of Waste (ATW) has rekindled.[29,32,33] Early ATW concepts[32] focused on strong thermalization of the energetic spallation neutrons generated by the interaction of a high-energy (~GeV) proton beam with a heavy-metal (W, Pb, U) target in order to transmute in a range of neutron energies where neutron absorption cross sections are high and inventories can be held to low values. More recent (conceptual) design directions for these ATW systems[29] are pointed towards the improved neutron economies associated with harder neutron spectra, and correspondingly would operate with higher inventories. These more recent ATW concepts are centered around pyrochemical processes of the kind originally developed for the IFR[34] and more recently elaborated[29,35] in the context of an ATW system designed to deal with the cumulating inventories of spent commercial nuclear fuels.

The Radkowsky Thorium Reactor[36] (RTR) portends the use of conventional LWR technology in a system that requires only thorium and low-enriched uranium (LEU, <20% ^{235}U) to achieve a power system with significantly reduced proliferation risk and increased access to the full and substantial thorium resource. The RTR uses fuel assemblies in a conventional LWR (PWRs or VVERs) based on a (supercritical) "seed" (enriched uranium) region surrounded by and driving a (subcritical) "blanket" (^{233}U breeding/fissioning thorium). The "seed" efficiently uses the uranium-based (<20% ^{235}U) driver with reduced plutonium generation (minimize natural uranium requirement), and the blanket efficiently generates *in situ* fissioning of ^{233}U to maximize power generation. The justification of the RTR "seed-in-blanket" fuel assemblies includes: a) significant reduction (or possible elimination) of fuel-cycle proliferation risk; b) reduced spent-fuel storage and disposal requirements; c) fuel-cycle cost savings; d) use of new fuel in existing reactors [PWRs or VVERs, with only minor hardware changes (fuel assemblies, control rods, burnable poisons); e) preservation of existing and proven safety and operational parameters; and f) use of existing fuel (but, not necessarily commercial) technologies]. These attributes are presently being subjected to detailed assessment.

Six potential (and incomplete) approaches to a nuclear future have been described: a) Once-Through LWRs (OT/LWR); b) MOX recycle in LWRs (MOX/LWRs); c) actinide (and possibly LLFP) destruction in ALMR/IFRs; d) IFR-based Self-Consistent Nuclear Energy Systems (SCNES) that permit no actinide or LLFPs to leave the reactor site; e) Accelerator-Driven Systems for actinide fissioning and LLFP transmutation (ADSs or ATWs); and f) enriched-uranium (20%) driven thorium blankets assembled into existing LWRs, (Radkowsky Thorium Reactor, RTRs). These partial, incomplete, and disparate concepts, however, with the exception of the stand-alone OT/LWRs and MOX/LWRs, have the following common goals:

- significantly reduce (nearly eliminate) present stocks of separated plutonium through the generation of energy therefrom, and to prevent the accumulation of future stocks of separated plutonium;
- keep all but operationally necessary MOX inventories in strong intrinsic (fission products) and protecting radiation fields during all fuel cycle operations;
- reduce or eliminate the flow of:
  - all actinides from the fuel cycle to the repository;
  - all LLFP from the fuel cycle to the repository;

These goals represent strategic elements of an architecture that offers a means to bridge to the sustainable future for nuclear power depicted in Fig. 10. This Integrated Actinide Conversion System (IACS)[37] is described in Fig. 13 in terms of five essential components required to bridge an ALWR/ELWR-based economy into a sustainable, plutonium- and/or uranium-breeding future where no idle and/or inherently unprotected plutonium inventories in any form exists. The IACS is a means (*e.g.*, an architecture) to this end, and a possible roadmap along with sign posts along the way are illustrated in Fig. 14 in terms of progress in each of the five component areas identified on Fig. 13: 1) elimination of separated plutonium in all forms; 2) improved ALWR/ELWR safety and economics; 3) discharge all nuclear fuels into an IMRSS; 4) enlist all actinides in the generation of energy; and 5) create a repository that is essentially free of actinides and LLFP could be any one of the concepts described above.

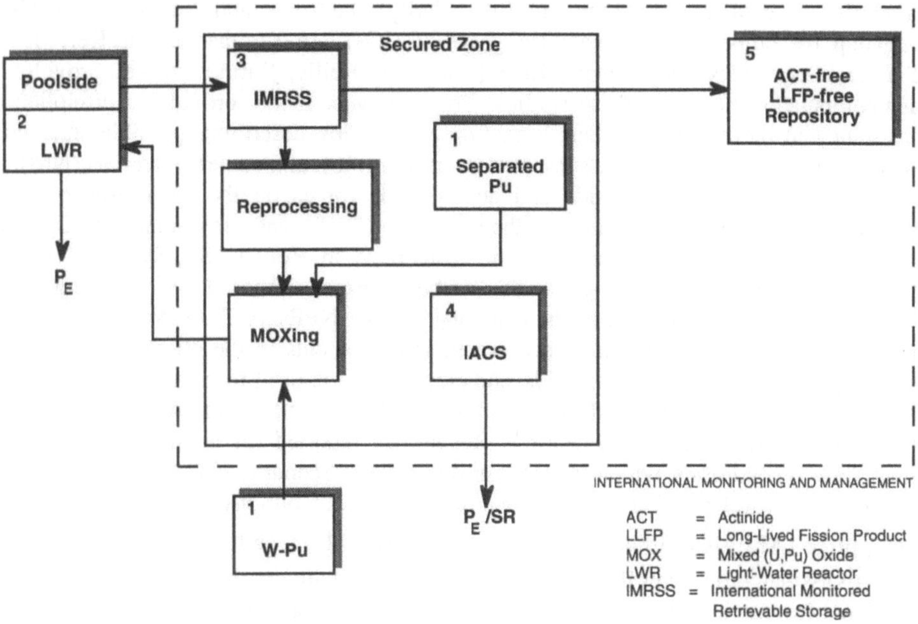

Figure 13. Integrated Actinide Conversion System (IACS) architecture.[37]

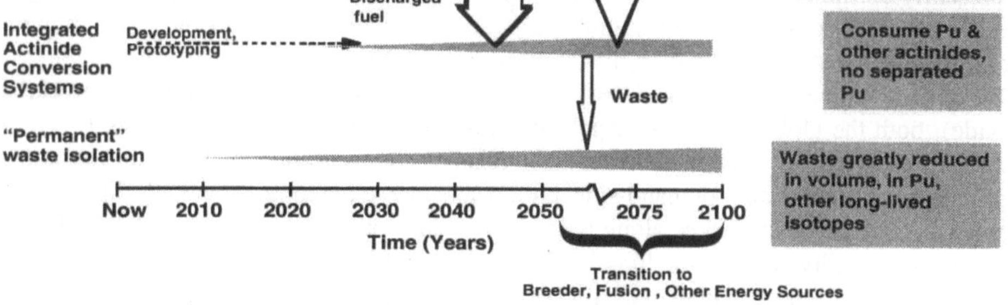

Figure 14. "Top-level" strategy for securing all nuclear materials and protecting the environment.[37]

IMPACTS/TRADE-OFFS

The results of impact and trade-off studies[38-41] examining (primarily) the flows and inventories of civilian plutonium and (secondarily) the roles and limitations of NE in mitigating the emission of greenhouse gases (GHGs) are summarized in Ref. 1. The behavioral economics ("top-down") ERB (Edmonds, Reilly, Barns)[42] model provides the main computational tool for these studies. This computer model provides a discipline and consistency to the creation of long-term future scenarios describing possible interactions between regional and global E^3 issues, nuclear energy (NE), and nuclear materials (NM).

The influences of supply-side and demand-side forces on the application of NE to regional and global energy mixes, and the economic (GNP) and environmental (GHG and NM accumulations) impacts that result are elaborated in Ref. 1. Supply-side forces are simulated through the surrogate of carbon taxes imposed at varying rates, CTAX($/tonneC/15yr), as well as the unit total cost of NPPs [UTC($/We)]. Demand-side forces are modeled through non-priced-driven improvements in the efficiency with which secondary energy (gases, liquids, solids, and electricities) is used to provide energy services to the residential/commercial, industrial, and transportation sectors - the Autonomous Energy Efficiency Improvement (AEEI) factor.[43] Results are reported for nominally "business-as-usual" (BAU) conditions, which are expressed as a Basis Scenario. The Basis Scenario used assumes: a) a world population in the year 2100 of ~11.7 billions; b) no carbon taxation; c) non-price (AEEI-like) improvements of ε_k = 0.0100 1/yr; and d) an annual increase in productivity that varies over the range 0.3-2.0 %/yr, depending on time and region. Many of these input assumptions have been subject to extensive parametric sensitivity analyses,[38,40,44] some of which are described in Ref. 1, which in addition to CTAX and AEEI variations reports on the sensitivities of NEs role in mitigating greenhouse gas emission to carbon taxation mode (*e.g.*, coupling back of carbon-tax-derived revenues to the rsegional economies *vis á vis* GDPs) and the cost of NPPs. The impact of carbon-dioxide emissions is monitored through estimates of both atmospheric carbon accumulations, W(GtonneC), and the increase in average global surface temperature, ΔT(K), that might result; a simplified integral-response (ocean-atmosphere) formulation is used to estimate ΔT.[45] These consequences of fossil-fuel (FE) use are *relative* because ΔT(K) is not translated into actual economic consequences, as reflected (for example) in possible GDP impacts[47,50]. The accumulation of plutonium in a range of forms for the Basis Scenario fuel cycle assumed (*e.g.*, phase-in of MOX recycle to 30% of all LWR reactor cores) is also expressed in terms of an inventory-based (temporally, regionally, form, and inventory dependent) proliferation-risk index,[46] PRI, which is also a *relative* indicator of possible consequences. Furthermore, as presently formulated, PRI monitors only proliferation risk from civilian plutonium and does not include possibly more significant risks associated with advances in uranium-enrichment technologies or the theft/diversion of NW components.

For the Basis Scenario considered (*e.g.*, no net sinks for either plutonium or carbon dioxide), both the global-climate-change (GCC) and the proliferation-risk indices (*e.g.*, ΔT and PRI, respectively) grow with time; only relative changes are monitored as the above-described supply-side and demand-side parameters are exogenously varied. The relative changes in GDP are also tracked along with changes in supply-side and demand-side drivers, giving a opportunity to monitor three-way (E^3) trade offs. Trade-off between supply-side (CTAX) drivers and demand-side (ε_k) drivers in mitigating the impacts of global warming, as computed from accumulations of atmospheric CO_2 in the form of the average global surface temperature response, ΔT(K), using the integral-response function reported in Ref.-45 are reported.[1] Including the Basis Scenario (CTAX = 0, ε_k = 0.0100 1/yr), the results for CTAX-ε_k combinations are also reported in Ref. 1 in terms of PE and NE demands, plutonium accumulations in a range of forms, and CO_2 emission rates and atmospheric accumulations.

This section reports only a sampling of the results reported in Ref. 1, which gives extensive graphic results from the parametric studies of supply-side, demand-side, and combined impacts on NE demand, GHG emissions, and GDP impacts. The primary-energy demand and mix for the Basis Scenario is given at a global level in Fig. 15, and the nuclear-energy component is given in Fig. 16, again at the global level. These aggregated results are compared to the variants reported as part of the Ref.-17 study. For this Basis Scenario, China becomes a major consumer of nuclear energy in the second half of the 21st Century. The magnitude and chronology of this dominance in NE demand depends sensitively on the assumed (regional) growths in economic productivity (GDP). Reference 1 gives the evolution of total civil plutonium for the Basis Scenario condition that assumes plutonium recycle in a (volume) fraction $f^f_{MOX} = 0.30$ of all LWR cores in each of thirteen global regions; the time evolution of CO_2 emissions, integrated emissions, and atmospheric accumulations, W(GtonneC), along with the average increase in global surface temperature, $\Delta T(K)$, are also reported in Ref. 1. The CO_2 emissions from each of the thirteen global regions show China becoming a dominant contributor to GHGs by 2025, in spite of its (world) dominance in nuclear energy shortly thereafter. The growth of (world) plutonium inventories with increasing atmospheric CO_2 inventories is correlated in Fig. 17. These plutonium inventories are divided into four categories: plutonium in reactor cores (REA); plutonium in spent fuel that can be recycled to LWRs (ACC); plutonium in spent fuel that can no longer be recycle to LWRs (REC, *e.g.*, have exceeded $N_{CYC} = 3$ recycles); and plutonium in separated form (SEP, *e.g.*, in processing or in fresh fuel). These inventories are used to estimate a (multi-attribute) proliferation-risk utility function, and, when discounted over time, a proliferation-risk index, PRI,[46] results. As noted previously, PRI is a monitor of proliferation risk associated only with with civilian plutonium and, therefore, is only an incomplete metric.

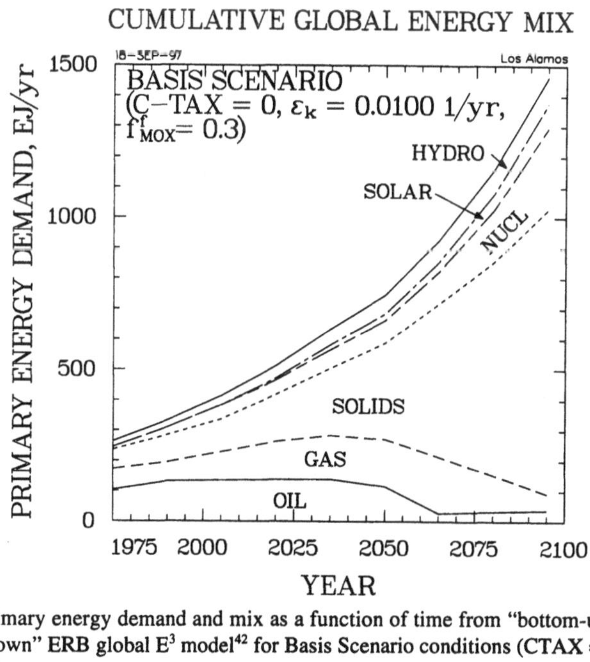

Figure 15. Global primary energy demand and mix as a function of time from "bottom-up" nuclear energy model[40] use in "top-down" ERB global E^3 model[42] for Basis Scenario conditions (CTAX = 0, $\varepsilon_k = 0.0100$ 1/yr, $f^f_{MOX} = 0.30$)[1,40]

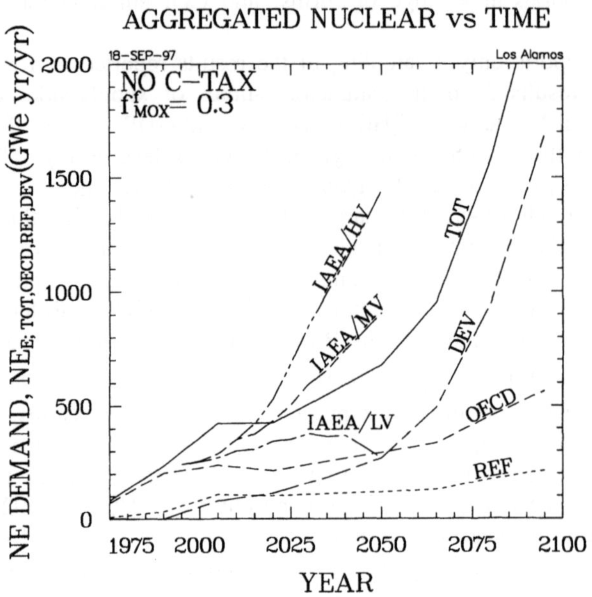

Figure 16. Nuclear energy demand from for Basis Scenario conditions (CTAX = 0, ε_k = 0.0100 1/yr, $f^f_{MOX} = 0.30$)[1,40]

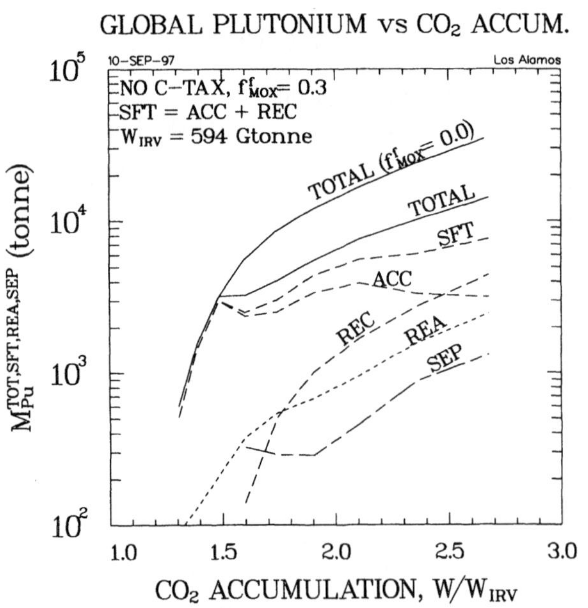

Figure 17. Cross-correlation of growing accumulations of global plutonium and atmospheric CO_2 for the Basis Scenario; growth in the four forms of plutonium are also shown [ACC = LWR-recyclable, REC = fully recycled and of use only in a fast-spectrum system (LMR[28] or ATW[29]), REA = reactor inventory; SEP = separated in processing or fuel fabrication, SFT = ACC + REC]; also shown is the buildup of total plutonium for the once-through LWR case, OT/LWR ($f^f_{MOX} = 0.0$).

Correlating the results of computed CO_2 emissions through an atmosphere-ocean model,[45] an estimate of the average global surface temperature rise, $\Delta T(K)$, is made. Figure 18 correlates PRI and ΔT with and without plutonium recycle; expectedly, both *relative* metrics increase with growth in FE and NE demand, with relative changes having most relevance. It is noted that the slightly increased cost of NE associated with plutonium recycle for the Basis Scenarios conditions leads to a slightly reduced market share for NE and a corresponding increase in the use of FE and slight increase in $\Delta T(K)$ by the year 2095. These trade-offs between plutonium inventories (or PRI) and atmospheric CO_2 inventories (or ΔT) are subject to detailed investigation in Ref. 1 under conditions where supply-side (CTAX) or demand-side (ε_k) forces are applied to shift the Basis Scenario; a sampling of these results is given below. The impacts of supply-side, demand-side, and fuel-cycle parameters on uranium-resource depletion are also reported in Ref. 1.

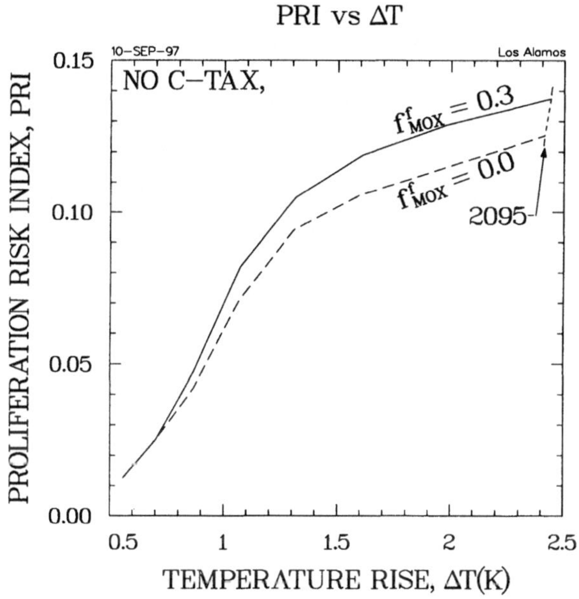

Figure 18. Relationship between proliferation-risk metric, PRI[46] (a weighted sum of inventory-based, discounted utilities), and estimated[45] increases in average global surface temperature, for the Basis Scenario ($f^f_{MOX} = 0.30$) and a once-through ($f^f_{MOX} = 0.0$) case.

For the series of demand-side cases studied, a carbon tax was applied at a rate CTAX($/tonneC/15yr) that starts in the year 2005. A slight reduction in PE demand accompanies the increases in carbon taxes, and a significant increase in demand for non-carbon energy sources [both NE and solar (PV)] accompanies the tax-induced decrease in FE demand. While the more expensive FE that results shifts the primary-energy mix to reduced- and non-carbon energy sources, the overall demand for energy is also reduced through the decreased income (GNP) brought about by higher overall energy prices. Figure 19 correlates the relative decrease in CO_2 emissions with the specific level of taxation; also shown is the result of a regression analysis of a number of earlier studies of this kind.[47] For these cases, the revenues collected as carbon taxes are not re-introduced into the economy. If this annual revenue is returned (*e.g.*, add to regional GDPs) with a multiplier M_{TAX} to compensate reductions in GNP caused by higher energy prices, consumption and CO_2 emissions increase, as is reported in Ref. 1.

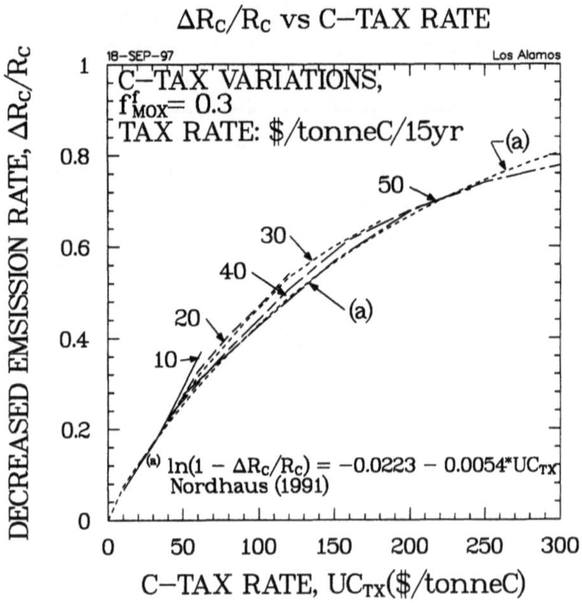

Figure 19. Relative decrease in carbon dioxide emission rates, $\Delta R/R_{CO_2}$, as a function of (linearly increased) unit carbon tax, UC_{TX}($/tonneC), and comparison with least-squares fit to a number of early assessments.[47]

Figure 20 combines positive (reduced ΔT) and negative (increased PRI) relative impacts of carbon taxes, showing the relative changes in each as carbon taxation is increased. Ultimately, these relative metrics must be extended to a common (absolute) basis of comparison (*e.g.*, GDP impacts), and in turn compared with other macroeconomic impacts of more expensive energy for a given tax-mode or revenue-recycle scenario. Through the behavioral-economic "lens" of the ERB model, the relationship between ΔT(K) and nuclear-energy demand is displayed on Fig. 21. The impact of these carbon-tax-driven increases in NE demand on the cumulative uranium requirement for this $f^f_{MOX} = 0.30$ case is reported in Ref. 1. For the CO_2-emissions stabilizing carbon tax rate of CTAX = 30 $/tonneC/15yr, the required uranium approaches that of the Total Conventional Resource of 15.5 Mtonne. Under these conditions and for the uranium-ore costing algorithm used,[40] a slight decrease in the cost of the LMR (LMR/LWR total capital cost ratio = 1.5 for these computations) would allow (economic) introduction of the breeder in the last quarter of the 21st century.

Like most of these E^3 models, the ERB model assumes an exogenous value for the non-price improvement of the efficiency with which secondary energy is converted to energy services over time. For the Basis Scenario, this AEEI-like parameter, ε_k, is assumed to be 0.0100 1/yr for all regions after the year 2005. The impact of increasing and decreasing ε_k uniformly for all regions on primary- and nuclear-energy demands is incorporated in Fig. 21. The evolution of the primary-energy intensity (annual primary-energy demand *per* GNP, MJ/$) and the impacts on CO_2 emissions and global warming are reported in Ref. 1.

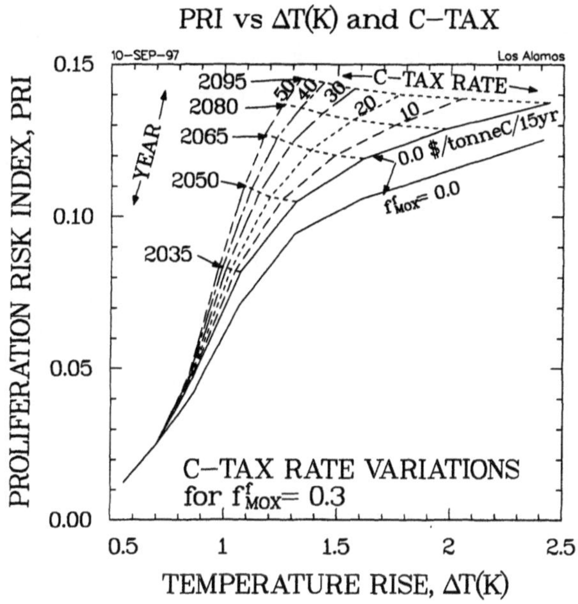

Figure 20. Relationship between evolution of proliferation risk index, PRI, and global temperature rise, ΔT(K), with plutonium recycle in LWRs and for a range of (linear) carbon taxation rates, CTAX($/tonneC/15yr).

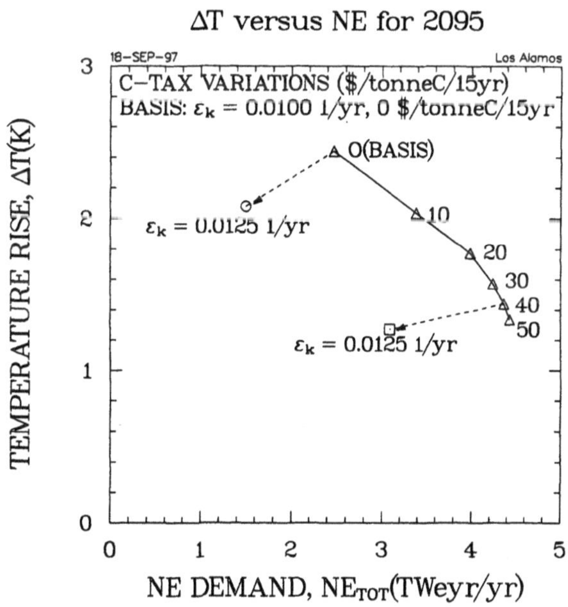

Figure 21. Impact of carbon-tax-induced increase in nuclear energy, NE_{TOT}, on reduced (final-year, 2095) average global surface temperature rise, ΔT(K); impact of increasing AEEI (Autonomous Energy Efficiency Improvement) parameter, ε_k(1/yr), from Basis Scenario value (ε_k = 0.0100 1/yr) is also shown; carbon taxes are applied linearly at the rates indicated, starting in 2005.

The parameter ε_k, however, should be tied to energy prices or cumulative experience in a way that its variation becomes endogenous to the model.[48] Generally, Fig. 21 indicates an approximate trade-off between supply-side (CTAX) drivers and demand-side (ε_k) drivers in mitigating potential impacts of global warming, as computed from accumulations of atmospheric CO_2 in the form of the average global surface temperature response, $\Delta T(K)$.

The cost of electricity computed by the modified ("bottom-up") NE algorithm used in the ("top-down") ERB model[40] includes both plant capital costs as well as O&M costs associated with key elements of the nuclear fuel cycle. The former is aggregated in terms of a unit total cost, UTC($/We), which varies both regionally and temporally. For the Basis Scenario, UTC is assumed to increase for all regions to an asymptotic value of ~2.4 $/We. Plant efficiencies and availabilities are also increased in according to the algorithms used in Ref.-17. A series of parametric studies were conducted[49] by driving UTC either upward or downward in order to affect a series of market-driven NE scenarios using UTC as an economic instrument. Specifically, transitions between the kinds of "continued-growth" and "phase-out scenarios" suggested in Ref. 24 (Fig. 9) were induced by applying the following time-dependent factor to the UTC cost instrument:

$$f_{UTC} = e^{-(t-t_o)/T_{UTC}} + \frac{1-e^{-(t-t_o)/T_{UTC}}}{1-e^{-(t_f-t_o)/T_{UTC}}}\left[f_{UTC}^{f} - e^{-(t_f-t_o)/T_{UTC}}\right]. \quad (1)$$

The modification of the Basis Scenario UTC for each region begins at time t_o (1990), the time t_f (2095) is the last time frame examined, and the exponentiation time is T_{UTC} (40 yr); all are correspondingly functions of region, as is the final factor by which UTC is increased(decreased), f_{UTC}^{f}. For the results presented herein, however, all global regions are treated uniformly. The results of the parametric study reported herein are base on variations of only f_{UTC}^{f} over the range [0.3,4.0]. To provide increased fitting flexibility, a two-phased version of this driver function was also used.[49]

The impact on NE demand of increasing or decreasing UTC for NE through the algorithm described above is reported on Fig. 22. Shown also on these figures are three variants suggested in a recent IAEA study[17] (Fig. 8), as well as the scenarios considered in the NEA/OECD study[24] (Fig. 9). By plotting the NE market share fraction *versus* f_{UTC}^{f} on log-log coordinates, some measure of NE "price elasticity" derive. While some of the observed behavior[1] results from the particular form of the driving equation, the transition from strongly "inelastic" to strongly "elastic" behavior as f_{UTC}^{f} increases results from complex interactions among the price-structure built into the logit-paradigm share fractions used in the ERB model.[42] It is noted from Fig. 22 that even stronger increases in the cost of NE do not totally eliminate that option because of this logit-based share paradigm, combined with the generally strong growth in China's GNP (and, therefore, demand for energy) for the cases being examined here.

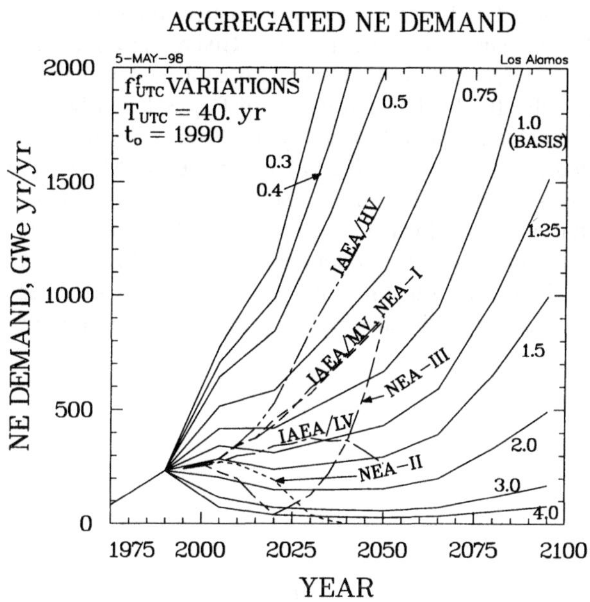

Figure 22. Impact of NE capital cost variations through the parameter f^f_{UTC} [Eq. (1)] on PE demand; f^f_{UTC} = 1. corresponds to the Basis Scenario; the high (HV), Medium (MV) and low (LV) variants from the recent IAEA study[17] (Fig. 8) is also shown, along with similar scenarios considered by a recent NEA/OECD study[24] (Fig. 9).

The impact of NE demand shifts, as driven by the (capital) cost instrument, on CO_2 emissions, accumulations, and potential for global warming is illustrated are reported in Ref. 1. The *per-capita* CO_2 emission rates tend to increase with time as economically advancing societies demand more-refined energy forms; both the *per-capita* carbon emission and the carbon intensity (kgC/GJ) rise significantly as the cost-driven reductions in NE share fraction is increased. Changes in the use of NE in the global energy mix also shift inventories of weapons-usable nuclear materials. The (plutonium) inventory-based metric of proliferation risk, PRI, is reported in Fig. 23 as a function of changing potential for global warming that accompany UTC-driven shifts in NE demand. This correlation is a counterpart of that shown on Fig. 20 for carbon-tax-driven increases in NE demand and the PRI *versus* ΔT trade offs related thereto. The relative impacts are clear: reduced (relative) proliferation risk through cost-induced reductions in NE demand increase the potential for global warming and *vice versa*. While the economic impact of cheaper or more expensive nuclear energy can be expressed in terms of changes in regional and global GNP, relating proliferation risk and/or global warming to a similar GNP basis is more difficult (precarious) and remains to be done.

Percentage changes in most of the key E^3 parameters reported up to this point as NE is made cheaper or more expensive through the cost algorithm represented above have been collected in the form of a sensitivity diagram on Fig. 24. All parameters give the percentage change reported in the year 2095 and are referenced to the f^f_{UTC} = 1.0 Basis Scenario. The change in atmospheric CO_2 inventories, ΔW_{CO_2} are express relative to the Basis Scenario the additions of atmospheric carbon since 1975 (when atmospheric inventories were 782 GtonneC, which increased by 1590 GtonneC for the Basis Scenario by 2095). The present value of world aggregated GNP, again expressed relative to the Basis Scenario, is included in Fig. 24 as ΔGNP(PV) using a 4% discount rate. Given that a mechanism is not specified for either the increases or decreases in UTC as f_{UTC} is exogenously varied, values of $f_{UTC} < 1$ can be viewed as a "subsidy" for NE and values of $f_{UTC} > 1$ can be considered as "excess revenues"; the present values of these "subsidies" or "excess revenues" have been computed and, when subtracted or added to ΔGNP(PV) gives the $\Delta GNP^{*}(PV)$ variable plotted on Fig. 24. In magnitude, these GNP shifts amount to fractions of a percent and are small compared to potential costs of global warming,[47,50] or the cost of direct carbon taxes to induce the use of reduced-carbon energy sources by making fossil fuels more expensive.[40] Figure 24 also gives the relative (again, to the Basis Scenario) percentage variations of f_{NE}, atmospheric CO_2 accumulations, global-warming potential, and proliferation-risk potential.

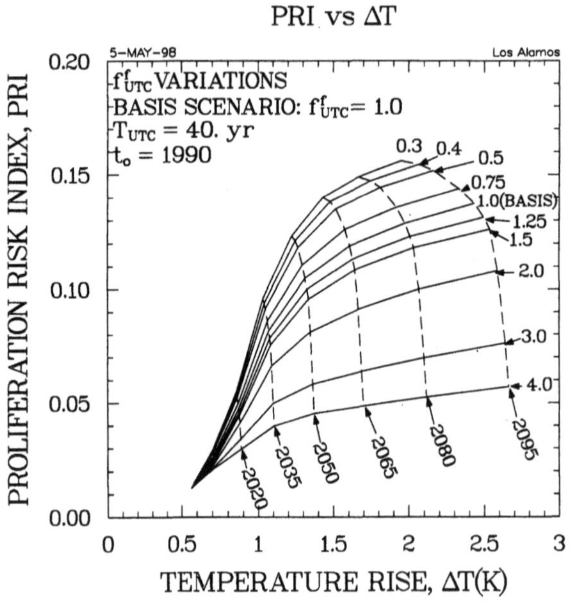

Figure 23. Relationship between the Proliferation Risk Index, PRI,[46] and the a measure of global warming potential as NE capital costs are varied through the parameter f^f_{UTC} [Eq. (1)]; isoclines are indicated, and f^f_{UTC} = 1. corresponds to the Basis Scenario.

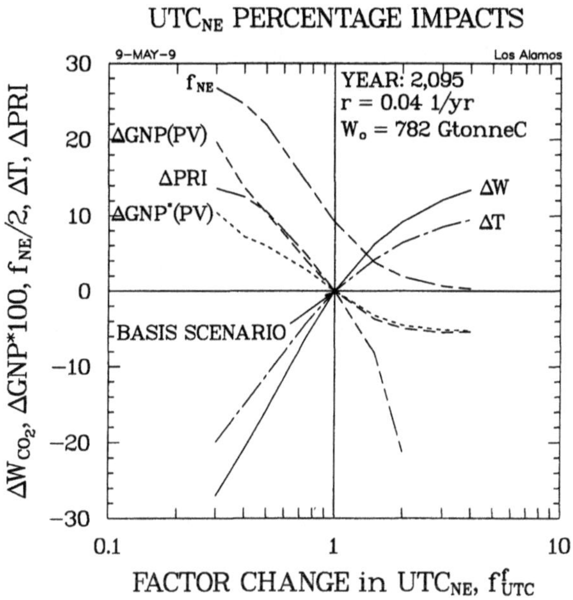

Figure 24. Summary of relative sensitivities of key metrics to NE capital-cost variations, as induced through the parameter f^f_{UTC} [Eq. (1)], measure in 2095 and referenced to the Basis Scenario. The following definitions and conditions are used: ΔW_{CO_2} is a relative atmospheric CO_2 inventory increase in the year 2095 (1,590 GtonneC for Basis Scenario, $f^f_{UTC} = 1.0$) compared to 1975 (782 GtonneC); ΔGNP(PV) is the percentage change in present value (r = 0.04 1/yr) of (world-aggregated) GNP; ΔGNP*(PV) is percentage GNP change corrected for the present value of any "subsidy" (if $f^f_{UTC} < 1$) or "over-charge" (if $f^f_{UTC} > 1$); f_{NE} is the fraction of all PE demand that is provided by NE, ΔT is the relative change in global average surface temperature relative to the Basis Scenario; and ΔW is the relative change in atmospheric CO_2 inventory relative to the Basis Scenario.

SUMMARY CONCLUSIONS

To varying degrees, under varying priorities, and depending strongly on country/regional cultural differences,[51] the advancement of nuclear energy must deal with four cardinal issues: *waste, proliferation, cost, and safety*. The degree to which sustainability can be achieved depends crucially on the resolution of these issues. While solutions are usually offered within a technolo-economic framework, the future of nuclear energy is being determined by these issues largely through a socio-cultural paradigm and a dwindling public acceptance of this technology. From a general perspective of energy consumption in terms of minimal requirements, needs, and constraints, a reduction of the enormous differences in *per-capita* primary-energy consumption that presently exists across the globe must be accompanied by a reassessment of how far above minimal levels (40-50 GJ/yr/*capita*) does "the good life" reside; bringing the entire world population of a world ~ 50 years hence to *per-capita* primary energy consumption levels of current affluence in North America is not feasible, even with large advances in technological efficiencies whereby primary and secondary energies are transformed into final-energy or end-use services.

Although increased efficiency through improved demand-side choices and advanced supply-side technologies will always be needed and desirable, new and expanded non-carbon energy sources like nuclear and solar energies will be required. The worldwide status attained by nuclear energy over the last three decades reflects a significant organizational and technological achievement, but reductions in NPP demand and deployment in developed countries at a time of growing opportunities in developing countries generates concerns related to the ability to sustain the complex infrastructure required for safe implementation and operations. In view of the complexity of this technology, the status of nuclear energy represents and enormous accomplishment, but uncertainties driven by a shifting customer (cultural) base, market transformation, and diminished public acceptance cloud longer-term global prospects. Key elements of the public acceptance issue are described, and pathways are suggested to improve acceptance of this technology. A range of technical approaches to the future of nuclear energy offer new directions that are key to improved public acceptance. New economic questions are arising with respect to financing and market forces for increased "efficiency" related to the increased globalization of capital markets in general, and the way in which each country is dealing with these growing global economic forces.[52] Finally, the long-term and global impacts and trade-offs for nuclear energy as viewed through the lens of a simplified energy-economics-environmental model emphasize how the future for nuclear energy may be tied to present concerns over the growing use of fossil fuels and possible climatic impacts related thereto.

Key findings emerging from this synthesis and analysis, as summarized above, are listed below:

- The electric supply industry is the only market for civilian nuclear energy and in this regard the nuclear industry is "on tap, but not on top";[21] however, this situation need not remain so in the future, as (if) NE enters the non-electric market (*e.g.*, hydrogen production, coal gasification, desalination of sea water, *etc.*).
- Public response to nuclear energy is value-laded and cultural in context; this condition has far-reaching implications for efforts to win greater acceptance of this technology.[25]
- The forces shaping public attitudes towards nuclear power are social-cultural in nature, and are not (directly) resolved within a technological-economic paradigm; these forces are related to:[53]
  - no perceived urgency (for new electric generation capacity);
  - nuclear energy is perceived as more costly than alternatives;
  - concerns related to being insufficiently safe;
  - little trust in governmental or industrial advocates;
  - concerns about health effects of low-level radiation;
  - concerns that means for dealing with high-level radioactive waste do not exist;
  - concerns about proliferation of nuclear weapons through the civilian nuclear fuel cycle.
- While important in dealing with the four cardinal issues for nuclear energy (waste, proliferation, cost, safety), "re-engineering" of nuclear systems alone will be ineffective in recovering public acceptance of this technology; pathways for increased public acceptance include:[25]
  - demonstrated record of safe operation of all nuclear facilities;
  - containment of catastrophic risk potential;
  - total separation of nuclear energy from nuclear weapons;
  - re-discover the benefits of nuclear energy;
  - deal satisfactorily with the waste;
  - re-establish fair, equitable, open institutions;
  - to the extent necessary for opening the above pathways, break with the past.
- Bridging to a nuclear-energy future requires that:[27]

- existing NPPs be secured through license renewal, continued reductions in O&M costs, and reduced demands on safety systems;
- evolutionary LWRs continue towards safer and more competitive systems;
- substantial progress be made on the technologies required to assure that prior to the year 2100:
  - -- all but the operationally minimal stocks of separated plutonium be eliminated;
  - -- all inventoried plutonium remains unseparated and isolated by a high radiation barrier;
  - -- all waste directed to repository be free of both actinides and long-lived fission products to the maximum extent practicable;
  - -- the world NPPs be operated with the minimum inventories of plutonium in all forms.

- Given that the LWR → ALWR/ELWR bridge (Fig. 10) to a viable nuclear-energy future can be established, a range of technologies remain to be explored and developed that assure: a) a fully minimize stockpiles of separated/accessible fissile material; and b) waste streams emanating from the NE fuel cycles of the future contain neither actinides nor long-lived fission products:
  - the stewardship philosophies embodied in the SCNES[28] or the IACS[37] concepts should be translated into technical realities;
  - the reality of any viable NE future will depend on limits to growth as established by: a) rate at which barriers to public acceptance of this technology is lowered; b) energy demand shifts and growths; c) economic (financing) limitations; and d) fuel resource limitations;
  - within the ground rules and reality checks listed above, the following approaches to a long-term NE future should be explored:
    - -- the long-term need for and economics of fissile-fuel breeders *versus* uranium-from-seawater/IACS (plutonium burning);
    - -- use of the thorium resource *vis á vis* the RTR;[36]
- Given that a bridge to a viable NE future cannot be constructed, the technological, (nuclear-materials) inventory, and overall infrastructural implications of a nuclear phase out[1,22,24] should be explored on both regional and global levels.
- Nuclear energy can make an important contribution to mitigating greenhouse gas emissions, but only by occupying market shares vacated by a more expensive (*e.g.*, directly or indirectly taxed) fossil fuel; stabilization to present CO_2 emission rates will require:[1]
  - NPP capacities of 4,500-5,000 GWe by the year 2100, corresponding to deployment rates of 80-90 GWe/yr after ~2030;
  - depending on uranium resource assumptions, breeder reactors will have to be deployed sometime around 2050 at a rate largely determined by the availability of startup plutonium;
  - equally substantial increases in renewable energy sources, particularly solar, will be needed.

NOMENCLATURE

ACC (tonnePu)	Accumulated LWR-recyclable plutonium
ACT	Actinide
ADS	Accelerator-Driven Systems
AEEI	Autonomous Energy Efficiency Improvement
ALMR	Advanced Liquid-Metal Reactor
ATW	Accelerator Transmutation of (nuclear) Waste
BAU	Business-As-Usual
BWR	Boiling Water Reactor
CFC	chlorofluorocarbons
CRIEPI	Central Research Institute for Electric Power Industry (Japan)
CTAX($/tonneC/15yr)	Carbon tax rate
CY	Calendar Year
DEV	Developing countries
DU	Depleted Uranium
E^3	Energy-Economics-Environment
EEU	Eastern Europe
ELWR	Evolutionary LWR
EN	Energy
ERB	Edmonds, Reilly, Barns model[42]
FBR	Fast Breeder Reactor
FE	Fossil Energy
FP	Fission Product
FSB	Fast-Spectrum Burner
f^f_{MOX}	final MOX core fraction
f^f_{UTC}	final factor change in UTC for nuclear energy
FSU	Former Soviet Union
GCC	Global Climate Change
GCR	Gas-Cooled Reactor
GDP(B$/yr)	Gross Domestic Product
GHG	Greenhouse Gas (CO_2, CH_4, N_2O, CFCs, *etc.*)
GNP(B$/yr)	Gross National Product
HLW	High-Level Waste
HV	High Variant
HVAC	Heating, Ventilation, Air Conditioning
IACS	Integrated Actinide Conversion System[37]
IAEA	International Atomic Energy Agency
IGCC	Integrated Gas Combined Cycle
IEA	International Energy Agency
IND	Industrialized countries
IRV	Industrial Revolution
LEU	Low-Enrichment Uranium (<20% ^{235}U)
IFR	Integral Fast Reactor
LGN	Liquefied Natural Gas
IIASA	International Institute for Applied Systems Analysis
LLFP	Long-Lived Fission Product
IIMRSS	International Interim Monitored Retrievable Surface Storage
IRV	Industrial revolution
LMR	Liquid-Metal (cooled) Reactor
LV	Low Variant
LWR	Light-Water (fission) Reactor
M^j_{Pu}	Plutonium inventory of jth form
M_{TAX}	Carbon tax multiplier if returned to the economy (GNP)

M_U(MtonneU)	Integrated natural-uranium resource usage
MCFC	Molten Carbonate Fuel Cell
ME	Middle East
MHTGR	Modular High-Temperature Gas (cooled) Reactor
MOX	Mixed (Plutonium, Uranium) Oxide Fission Fuel
MRS	Monitored Retrievable Storage
MV	Medium Variant
N_{CYC}	Number of MOX recycles in LWRs
NDT	Non-Destructive Testing
NE	Nuclear Energy
NEA	Nuclear Energy Agency
NM	Nuclear Materials
NPP	Nuclear Power Plant
NW	Nuclear Weapon
O&M	Operating and Maintenance (annual charges)
OECD	Organization for Economic Cooperation and Development
OT	Once-Through LWR
OTEC	Ocean Thermal Conversion Energy
PAFC	Phosphoric Acid Fuel Cycle
PE	Primary Energy
PHWR	Pressurized Heavy-Water Reactor
POP	Population
ppmv	Parts-per-million by volume
PRI	Proliferation Risk Index
PUREX	Plutonium-Uranium Recovery Extraction
PV	Photovolatic or Present Value
PWR	Pressurized-Water Reactor (LWR)
R&D	Research and Development
RE	Renewable Energy
REA (tonnePu)	Reactor plutonium
REC (tonnePu)	Non-recyclable (to LWRs) plutonium
RG	Reactor-Grade plutonium
RTP	Radkowsky Thorium Fuel Reactor[36]
RU	Recycle Uranium
SCNES	Self-Consistent Nuclear Energy Systems[28]
SE	Secondary Energy
SEP (tonnePu)	Separated (reprocessing, fabricated MOX) plutonium
SFT (tonnePu)	Total spent fuel plutonium
SR	Support Ratio (LWR/FSB)
UC_{TX} ($/tonneC)	Unit costs of carbon tax
UTC ($/We)	Unit Total Cost for nuclear energy
VVER	Russian PWR
W(GtonneC)	Atmospheric carbon dioxide inventory, ppmv = W/2.13
W_o(GtonneC)	Integrated carbon dioxide emissions
W-Pu	Weapons Plutonium
W_{IRV} (GtonneC)	Atomicspheric carbon inventory at time of industrial revolution
ΔT(K)	Global average surface temperature rise[45]
ε_k (1/yr)	Annual improvement in secondary energy → energy service conversion (AEEI-like parameter)

References

1. R. A. Krakowski, *Re-Engineering Fission: Reactors for Safe, Globally Sustainable, Proliferation-Resistant, and Cost-Effective Nuclear Power,* Los Alamos National Laboratory document LA-UR-98-3768 (August 25, 1998).
2. A. K. N. Reddy, R. H. Williams, T. B. Johansson, Energy after Rio: prospects and challenges, *United Nations Development Programme Report* (1997).
3. C. G. Darwin, *The Next Million Years*, Doubleday and Company, Inc. Garden City, NY (1953).
4. R. G. Watts (Ed.), *Engineering Response to Global Climate Change: Planning a Research and Development Agenda*, Lewis Publishers (1997).
5. W. Lutz (Ed.), *The Future Population of the World: What Can We Assume Today?,* International Institute for Applied Systems Analysis, Earthscan Publications Ltd., London, UK (1996).
6. S. Yoda, Nuclear Power in the 21st Century - The Challenge to Overcome the 'Trilemma', *Proc. Intern. Conf. on Future Nuclear Systems, Global'97, 1, 2,* Yokohama, Japan, (October 5-10, 1997).
7. International Energy Agency, *World Energy Outlook*, OECD, Paris (1996).
8. International Atomic Energy Agency, Energy, electric, and nuclear power estimates for the period up to 2015, *Reference Data Series No. 1* (July 1997 Edition), Vienna (July 1997).
9. G. H. Steven, Role of Nuclear Power in Sustainable Energy Systems, *Proc. Intern. Conf. on Future Nuclear Systems, Global'97, 1, 118,* Yokohama, Japan, (October 5-10, 1997).
10. Y. Uchiyama, *Life Cycle Analysis Of Power Generation Systems*, Central Research Institute of Electric Power Industry Report (CRIEPI, Japan) (1995).
11. V. Smil, *General Energetics: Energy in the Biosphere and Civilizations*, John Wiley and Sons, New York, NY (1991).
12. United Nations Organization, *Yearbook of World Energy Statistics*, UNO, New York, NY (1980).
13. World Bank, *World Development Report*, Washington, DC (1988).
14. J. Sheffield, World population growth and the role of annual energy use per Capita, *Technology Forecasting and Social Change*, 59, 55 (1998).
15. World Energy Council, *Energy for Tomorrow's World - the Realities, the Real Options and the Agenda for Achievement*, St. Martin's Press, Kogan Page Ltd., London, (1993).
16. D. Meneley (Chm.), Key Issues Paper No. 3: future fuel cycle and reactor strategies, *International Symposium on Nuclear Fuel Cycle and Reactor Strategies: Adjusting to New Realities*, Vienna, Austria (June 3-6, 1997), International Atomic Energy Agency, Vienna (1997).
17. H. F. Wagner (Chm.), Key Issues Paper No. 1: global energy outlook, *International Symposium on Nuclear Fuel Cycle and Reactor Strategies: Adjusting to New Realities*, Vienna, Austria (June 3-6, 1997),International Atomic Energy Agency, Vienna (1997).
18. J. Lochard (Chm.), Key Issues Paper No. 4: safety, health, and environment implications of the different fuel cycles, *International Symposium on Nuclear Fuel Cycle and Reactor Strategies: Adjusting to New Realities*, Vienna, Austria (June 3-6, 1997), International Atomic Energy Agency, Vienna (1997).
19. Projected costs of generating electricity: update 1997, *OECD/NEA-IEA Report* (1998).
20. N. J. Diaz, Nuclear technology: global accomplishments and opportunities, *Nuclear News*, 36 (May 1998).
21. *Uranium 1995: Resources, Production, and Demand, Organization of Economic Cooperation and Development (OECD),* Nuclear Energy Agency (NEA), Paris (1996).
22. P. Beck, *Prospects and Strategy for Nuclear Power: Global Boon or Dangerous Diversion?,* Earthscan Publications, Inc. (1994).
23. N. Nakicenovic (Study Director), Global energy perspectives to 2050 and beyond, *International Institute for Applied Systems Analysis (IIASA) and World Energy Council (WEC) Report* (1995).
24. Nuclear power and climate change, *Nuclear Energy Agency (NEA) / Organization for Economic Cooperation Development (OECD) Report* (April 1998).

25. R. Kasperson, Can nuclear power gain public acceptance?, *Massachusetts Institute of Technology Report MIT-ANP-CP-002*, p. 3-2, Proc. 2nd Intern. Conf. on the Next Generation of Nuclear Power Technology, (October 25-26, 1993).

26. *Nuclear Energy: Perception Gap, Nuclear Energy Insight, Nuclear Energy Institute*, p. 8, (March 1998).

27. N. E. Todreas, What should our future nuclear energy strategy be?, *Massachusetts Institute of Technology report MIT-ANP-CP-002*, p. 7-2, Proc. 2nd Intern. Conf. on the Next Generation of Nuclear Power Technology, (October 25-26, 1993).

28. N. Takagi, R. Takagi, and H. Sekimoto, Effect of decontamination factor of recycled actinide and FP on the characteristics of SCNES, *Prog. in Nuclear Energy*, 32(3/4), 441 (1998).

29. F. Venneri, Ning Li, M. Williamson, M. Houts, and G. Lawrence, Disposition of nuclear waste using subcritical accelerator-driven systems: technology choices and implementation scenario, *Proc. 6th Intern. Conf. on Nuclear Engineering (ICONE-6)*, ASME Publication (May 10-15, 1998).

30. M. Steinberg, G. Wotzak, and B. Manowitz, *Neutron Burning Of Long-Lived Fission Products For Waste Disposal*, Brookhaven National Laboratory Report BNL-8558 (1958)

31. G. P. Lawrence, R. A. Jameson, and S. O. Schriber, Accelerator technology for los alamos nuclear-waste transmutation and energy-production concepts, *Proc. Intern. on Emerging Nuclear Energy Systems*, Monterey, CA (June 16-21, 1991)

32. W. C. Sailor, C. A. Beard, F. Venneri, and J. W. Davidson, Comparison of accelerator-based with reactor-based nuclear waste transmutation schemes, *Prog. in Nuclear Energy*, 28(4), 359 (1994).

33. C. D. Bowman, E. D. Arthur, P. W. Lisowski, G. P Lawrence, R. J. Jensen, J. L. Anderson, et al., Nuclear energy generation and waste transmutation using an accelerator-driven intense thermal neutron source, *Nucl. Instr. and Meth.* A320, 336 (1992).

34. J. J. Laidler, J. E. Battles, W. E. Miller, and E. C. Gay, Development of IFR pyroprocessing technology, *Proc. Global '93 International Conf. on Future Nuclear Systems*, 2, 1061, Seattle, WA (1993)

35. M. A. Williamson, Chemistry technology base and fuel cycle of the los alamos accelerator-driven transmutation system, *Proc. Global '97 International Conf. on Future Nuclear Systems*, 1, 263, Yokohama, Japan (1997).

36. A. Galperin, P. Reichert, and A. Radkowsky, Thorium fuel for light water reactors-reducing proliferation potential of nuclear power fuel cycle, *Science and Global Security*, 6, 225 (1997).

37. E. D. Arthur, P. T. Cunningham, and R. L. Wagner, Jr., *An Architecture for Nuclear Energy in the 21st Century*, Los Alamos National Laboratory report LA-UR-98-1931 (June 29, 1998).

38. R. A. Krakowski, J. W. Davidson, C. G. Bathke, E. D. Arthur, and R. L. Wagner Jr., *Nuclear Energy and Materials in the 21st Century*, International Symp. on Nuclear Fuel Cycle and Reactor Strategy: Adjusting to New Realities, Vienna, Austria (June 3-6, 1997) IAEA Publication (1998).

39. R. A. Krakowski, J. W. Davidson, C. G. Bathke, E. D. Arthur, and R. L. Wagner Jr., Global economic/energy/environmental (E3) modeling of long-term nuclear futures, *Proc. Global '97 International Conf. on Future Nuclear Systems*, 2, 885, Yokohama, Japan (October 5-10, 1997).

40. R. A. Krakowski and C. G. Bathke, *Long-Term Nuclear Energy Strategies*, Los Alamos National Laboratory document LA-UR-97-3826 (September 24, 1997)].

41. R. A. Krakowski, *The Role of Nuclear Energy in Mitigating Greenhouse Warming*, International Conf. on Environment and Nuclear Energy, Washington DC (October 27-28, 1997), Plenum Press, New York, NY (1998).

42. J. Edmonds and J. M. Reilly, *Global Energy: Assessing the Future*, Oxford University Press, New York, NY (1985).

43. A. S. Manne and R. G. Richels, Buying Greenhouse Insurance: *The Economic Costs of Carbon Dioxide Emission Limits*, MIT Press, Cambridge, MA (1992).

44. R. A. Krakowski, *Energy-Economic-Environment (E^3) Modeling Activities/Capabilities at Los Alamos National Laboratory*, Los Alamos National Laboratory document LA-UR-98-945 (March 8, 1998).

45. K. Hasselmann, S. Hasselmann, R. Giering, and V. Ocanna, Optimization of CO2 emissions using coupled integral climate response and simplified cost models: a sensitivity study, climate change: integrating science, economics, and policy, N. Nakicenovic, W. D. Nordhaus, R. Richels, and F. L. Toth (Eds.), *International Institute for Applied Systems Analysis (IIASA) report CP-96-1* (1995).

46. R. A. Krakowski, *A Multi-Attribute Utility Approach to Generating Proliferation-Risk Metrics*, Los Alamos National Laboratory document LA-UR-96-3620 (October 11, 1996).

47. W. D. Nordhaus, The cost of slowing climate change: a survey, *The Energy Journal*, 12, 37 (1991).

48. S. Messner, *Endogenized Technological Learning in an Energy Systems Model*, International Institute for Applied Systems Analysis document RR-97-15 (November 1997).

49. R. A. Krakowski, *Preliminary Parametric Studies Using the ERB Model in Search of Demand Scenarios for Use by IAEA Consultancy E^3 Study Team*, Los Alamos National Laboratory document LA-UR-98-2252 (May 18, 1998)

50. R. Repetto and D. Austin, *The Costs of Climate Protection: A Guide for the Perplexed*, World Resources Institute report (1997).

51. G. Rochlin, Nuclear technology and social culture, *Massachusetts Institute of Technology report MIT-ANP-CP-002*, p. 7-11, Proc. 2nd Intern. Conf. on the Next Generation of Nuclear Power Technology, (October 25-26, 1993).

52. R. C. Longworth, *Global Squeeze: The Coming Crisis for First-World Nations*, Contemporary Books, Chicago, IL (1998).

53. J. F. Ahern (Chm.), *Nuclear Power: Technical and Institutional Options for the Future*, National Research Council, National Academy Press (1992).

THE GLOBAL ENERGY SITUATION IN THE NEXT CENTURY AND ITS IMPLICATIONS FOR NUCLEAR POWER

Peter Beck

The Royal Institute of International Affairs
10, St. James's Square
London SW1Y 4LE

INTRODUCTION

The nuclear industry expresses confidence (at least in public) that nuclear power will have a rosy and expanding future during the next century. Notwithstanding today's bleak situation, it believes that energy demand will continue to increase, that, owing to the problems of climate change, use of fossil fuel will have to be curtailed and that the price of oil and gas is likely to rise as more unconventional resources come into play.

Others, outside the industry, seem equally confident that nuclear power has no future and will be abandoned during the next century. They believe that nuclear energy is too dangerous and uneconomic, that it is secretive and that the public has no confidence in it so that governments, at least in democracies, are loath to assist the industry. Furthermore, they argue that the energy form will not be needed if sufficient effort is expended on improving energy efficiency and on the development of renewable energy.

Both parties argue that the other is either illogical or not truthful and is too biased to be worth talking to. As a result, there is no constructive contact between the two sides, with most meetings and conferences arguing the case the audience wishes to hear. Politically, the anti-nuclear lobbies seem to be gaining ground in a number of countries, but the pro-groups believe that this will change during the next energy crisis, which, from past experience, is bound to come sooner or later.

The present circumstances have led to a stand-off with hardly any decisions taken. The purpose of this paper is to assess this deadlock and consider ways of breaking it. The paper is in three sections:

> The first looks at the energy scene in the next century and tries to answer the question whether there is a strong case for either the pro- or the anti- nuclear view. If, on present information, there is no clear winner between the two, logic would imply that both options should be kept open.
>
> The second section considers what 'keeping the nuclear option open' implies and how it might be achieved in practice.
>
> Finally, some conclusions are reached from the two sections.

Preparing the Ground for Renewal of Nuclear Power, edited by
Kursunoglu *et al.* Kluwer Academic / Plenum Publishers, New York 1999.

ENERGY IN THE NEXT CENTURY.

The ability to forecast the future is not given to mankind. Ancient Greece knew that well: it needed the intervention of the gods to foretell the future. Experience, especially in the field of energy during this century have confirmed this inability; most forecasts, even if only for a few years ahead, proved spectacularly wrong. Consensus forecasts have a particularly bad track record. Some mavericks, often vilified by main-stream experts, did get it right, but as not all mavericks can be right, how does one choose the right one?

Such thoughts lead to the question whether it is worthwhile to discuss the development of the energy scene over the next fifty or even hundred years. The answer lies with the purpose of the exercise. If the purpose is numerical forecasts, the answer must be *no;* if it is to achieve a better understanding of the forces - technological, economic, political, social, etc. - which will mould the energy scene, the answer is *yes.*

The methodology used for such a study is the development of scenarios , i.e. of alternative paths along which the future could unfold. In these, widely different assumptions, such as about the speed of technical progress or World trade developments, can be tested for their effect on energy. Such understanding is of value in assessing longer term strategies and short term actions to ensure that such strategies can be implemented. Time in the energy field - between initial laboratory work and a significant effect in the market place - tend to be long, especially in nuclear power, where they have to be measured in decades.

For the purpose of this paper, use was made of scenarios drawn up by the World Energy Council and the International Institute for Applied Systems Analysis in 1995[i]. These were widely accepted as a basis for studies and both organisations are seen as neutral within the pro- and anti- nuclear field. Tables 1. to 3.have been derived from this reference.

To study future energy demand and supply, three families of scenarios were drawn up, each incorporating very different assumptions about economic and technological growth and about political stability, but all within the bounds of feasibility.

The basic assumptions in these scenarios are:

Scenario A, "High Growth": Assumes high economic and technological progress and a political climate favouring free trade and market forces. The scenario has three sub-cases, largely dealing with energy supply:

A1 presumes technologically successful developments of oil and gas resources in the first half of the century, with nuclear and renewables slowly taking some of the burden in the second.

A2 makes the assumption that the pressure for limiting CO_2 emissions abates, so that there can be increasing use of coal as oil and gas resources diminish.

A3 assumes technological developments in the field of nuclear energy and renewables to make it possible to reduce fossil fuel to a secondary role by the end of the century.

Scenario B, "Middle Course" : This is a 'business as usual' case, an extrapolation of today's trends. From today's standpoint this scenario is seen as rather likely, except for the experience that long term extrapolations cannot be trusted.

Scenario C, "Ecologically Driven": Here it is assumed that governments take policy decisions to accelerate improvements in energy efficiency and remove obstacles in the development and spread of renewable energies. This scenario has two sub-sets:

C1 assumes highly successful advances with renewables and energy efficiency, making it possible by the end of the century to reduce fossil fuel use to no more than a fifth of total energy demand, even though it also assumes the phase-out of nuclear energy.

C2 looks at a situation where nuclear energy has managed to resolve its problems and becomes the second largest energy source after renewables.

Table 1 The WEC/IIASA Scenarios.

Scenarios		A, High Growth		B, Middle Course		C, Ecologically Driven	
Energy Intensity Improvements		medium		low		high	
Fossil Fuel. Resource		high		medium		low	
Other		high		medium		high	
Environmental taxes		No		No		Yes	
Year	1990	2050	2100	2050	2100	2050	2100
Population x 10^9	5.3	10.1	11.7	10.1	11.7	10.1	11.7
Primary Energy TWy	12.9	35	63	28	49	20	29
% in OECD	47	27	18	28	16	21	11

Energy Demand.

The above Table shows the main assumptions in the scenarios and the resulting primary energy demand. Although the latter figures should be taken as order of magnitude only, they do show the vast differences in energy demand already by 2050, which arise from the different assumptions in the scenarios. The main determinants of future demand are *population growth, the level and type of economic growth and improvement in energy intensity (*energy use per unit of GDP). All three are driven by technological change as well as by changes in behaviour and politics and are subject to huge uncertainties.

As an example, the population data in the Table are based on UN median projection, which assume stabilisation in the 2nd half of the century. It is also assumed that most of the increase in population will be in developing countries and that, unless positive steps are taken by governments (Scenario C), most of the additional numbers will live in cities. If, indeed, some three quarters of the population will live in urban areas and many in mega-cities, provision of adequate energy to them could become a daunting task and affect the type of fuel used as well as the rate of improvement in energy efficiency. It should also be noted that work now being done for a fresh set of scenarios (likely to be published in 1999) may well have to include widely different population projections for different scenarios, especially for the time beyond 2050.

Improvements in energy efficiency is another source of uncertainty. In developed countries they could be greatly assisted by environmental taxes, as shown in some detailed studies on the US position[ii]. These show the possibility of a 10%+ saving already by 2010 compared to a 'business as usual' case. The question for developing countries is whether they will be able and willing to make use of the best available technology, or whether they will have to make do with second best.

Reduction of the present disparities in energy consumption (with some 20% of the richest population using some 55% of energy and 20% of the poorest using 5%) is assumed under all scenarios, but the presumed rate of change varies for different regions and by scenario. This is another source of considerable uncertainty.

The wide range of the demand projections are, thus, not surprising and may well not encompass the full range of possibilities.

Energy Supply.

Uncertainties are just as great in this area. Here the main determinants are *the resource base of the many alternatives and their relative attractiveness to individual countries from the point of view of economics, security and environmental concerns.* Again, technological change and politics will drive these factors and determine the balance between the various sources of energy.

The WEC/IIASA scenarios show such balances for the different scenarios; they are based on very many assumptions which were chosen to give a wide range of answers, all within the bounds of feasibility.

Table 2 shows the resource base for the main energy forms, with the energy mix under the various scenarios shown in Table 3.

Table 2. Estimates of Energy Resources[1], TWy.

	1990 - Consumption.	Reserves (1)	Resources (2)	Resource base (1+2)	Novel type resources
Oil, Conventional.	4.5	210	200	410	
Unconventional.		270	470	740	2600
Natural Gas, Conv	2.4	200	390	590	400
Unconv.		270	360	630	
Methane Hydrates					26000
Coal	3.1	850	3900	4750	4200
Total	10.0	1800	5320	7120	33200
Uranium[(a)]	0.7	80	280	360	200
with recycle[(b)]		4800	17000	21800	12000

(a) using Thermal Reactors only (b) including use of Fast Reactors

[1] Definitions: **Reserves** are those quantities which geological and engineering information indicate with reasonable certainty that they can be recovered from known reservoirs using known technology and under existing economic conditions.
Resources have less certain geological assurance and/or cannot be as yet extracted with present technological means under existing economic conditions.
As technology advances, reserves will increase at the expense of resources and this has been happening since the dawn of the oil industry. The figure for reserves is thus a very misleading measure of future availability of fossil fuel and, indeed, of Uranium also.
The Resource base is the sum of the two and a rather better guide.

Novel type of resources cover very speculative plays; methane hydrates or Uranium from seawater are examples. They may be sources for the future, but are not taken into account in the scenarios.

Table 3. The Energy Mix under different Scenarios for the year 2050.

Scenario	1990		A		B	C	
		A1	A2	A3		C1	C2
Primary Energy, TWy	12.9	35	35	35	28	20	20
Mix, %: Coal	23	15	32	9	21	11	10
Oil	36	32	19	18	20	19	18
Nat. Gas	18	19	22	32	23	27	24
Nuclear	5	12	4	11	14	4	12
New Renewables	2	16	17	24	15	29	26
Old Renewables [a]	16	6	6	6	7	10	10

[a] Includes non-commercial and hydroelectric.

With the exception of nuclear energy, which will be discussed in Section II of the paper, comments are made below about the individual fuels:

Oil. As seen from Table 2, the reserve/production (R/P) ratio for conventional oil in 1990 was around 45 years and it is about the same today. Fifteen years ago the R/P ratio was 32 years, whilst in the meanwhile, production has increased by some 10%. So far, therefore, proven reserves have increased at a greater rate than production. This was not only due to new finds, but also owing to considerable advances in technology and engineering organisation which had the effect of increasing the amount of economically attractive reserves in existing fields and bring into reserves fields which were previously uneconomic[iii]. Conventional oil production only recovers around 30% of oil in place. Advances with enhance oil recovery (EOR) have also contributed to the healthy state of reserves. The industry believes that such advances will continue, so that up to now there was little need to consider the more unconventional oils, such as heavy oil, tar sands and shale oil. Most of today's production costs are below $7/bbl, whilst unconventional sources may cost some $20 to 25[iv].

As regards price, so far in the history of oil there has hardly been any relationship between price and cost. Today's prices are set by global markets, but they are also manipulated through a political process. As long there is the perception of ample supplies, prices will stay weak; should there be a crisis, they will rocket up, but only to come down again when the crisis is over. It may well be that this situation will last for a number of decades, possibly even to mid-century. Significant and lasting price increases cannot be expected for some time.

If demand increases (Scenario A1), production from OPEC countries would become more important which might lead to higher prices. However, most of these countries have very high R/P ratios and with talk about phasing out oil during the century owing to the climate change issue, they may well be too concerned about being left with valueless oil in the ground, to repeat the policies of 1973.

There are voices who strongly disagree with the above analysis[v]. These believe that oil production will peak early in the next decade and fall quite rapidly thereafter. That would trigger an oil price shock which would change the energy scene permanently. The industry and most energy economists do not agree, but as mentioned earlier, some mavericks turn out to be right; can this be one of them?

Natural Gas. Outside N. America, natural gas has only stated to be important during the last 25 years. In that period production doubled and reserves increased threefold. Where available, it has become the preferred feedstock for power generation; it is seen as a clean fuel and it has the potential for becoming a transport fuel.

There are some problems to overcome. With most of the reserves far away from centres of consumption, gas projects of the future are likely to be large and costly. So far, however, technical and managerial improvements have managed to cope with reducing prices[vi], but will this continue? Also, Methane is a 'green-house gas'; it is, therefore important to limit leaks which can be significant in pipeline systems. Furthermore, some gas fields have gas with a high CO_2 content which has to be taken into account when considering greenhouse gas emissions. Nevertheless, the future for natural gas looks bright.

In addition to the resources of natural gas, there are vast deposits of Methane Hydrate, a crystalline material formed under certain conditions of pressure and temperature by methane and water. The U.S. Dept. of Energy estimates that U.S. Methane Hydrate deposits may hold some 200 times the amount of methane held in U.S. conventional fields.[vii] However, so far, no commercial use has been made of these deposits and there are indications that, compared to conventional gas, extraction could be difficult and costly.

Coal. The coal industry faces a very uncertain future for two reasons: firstly, its high CO_2 emission compared to oil and gas and secondly that it is dirty to handle and that it can be highly polluting, especially when used in older plant. Technological advances have made it possible to achieve 'clean combustion', but such plant is expensive and clean transport and handling is difficult. Although coal may become the fuel 'of last resort' in most countries, only to be used on a large scale if problems with other fuels are foreseen, it may well continue to be the main source of energy in countries such as China and India which have large coal deposits, but few resources of other fuels.

In the long term, coal gasification to augment natural gas may be possible in some areas, especially if natural gas development becomes less buoyant than now generally believed.

Renewables. This is a highly controversial subject. In distinction to nuclear energy, there is little opposition against renewables, but there are many sceptics who do not believe its ability to capture a significant proportion of the energy market as early as mid-next century. Those working in the field and much of the environmental movement believe that rapid progress *is* possible if adequate funds for development are available and if governments give renewables a helping hand when restructuring the electricity market. They see no other way to a 'sustainable' future.

Table 2 shows the proportion of new renewables in 1990 to be just 2%. These include a large number of technologies, the most important being *biomass, solar, wind, geothermal, ocean energy systems* and *small hydroelectric schemes,* each having many alternative ways of harnessing energy. Work on all these is being done in many countries, but in most cases technology has not yet reached commerciality.

Although legislation, favourable to renewables, was enacted in 1978, new renewables have, in 1996, only supplied some 3 ¾ % of U.S. energy and with low growth year-on-year[viii]. However, present work on the next generation of projects appears to indicate that costs may well on average be halved, that most would be commercially justifiable and that growth will, therefore accelerate[ix].

Sceptics are doubtful about such claims and argue that the whole development process for renewables as well as the incentives to install such capacity is driven by governments, which at the same time maintain that they wish to withdraw from active involvement in the energy market and leave its development to market forces. There is, thus, a clash between two aims of many governments: on the one hand they wish to have the benefits of competition in the electricity market, but on the other, they have committed themselves to be environmentally responsible and move towards sustainability.

In addition, sceptics point out that the capital cost of new schemes tends to be badly underestimated. It is, therefore, probable that the real cost of the many different renewables will not become clear until after the second or third generation of plants. This implies that it may be ten to fifteen years before we shall know which of the many renewables' technologies are likely to have a long-term future. The initial full scale projects will, therefore, be risky investments. Bearing in mind the aim of having in future a highly competitive electricity market, profits from even a successful investment will be low, unless somehow assisted by governments. But will they continue to do so and who will choose what to assist?

The sceptics raise other questions. Firstly, with a few exceptions, the scale of renewables' power stations will be small compared to present-day stations and will require large areas of land. That may be acceptable for the countryside, especially in many of the Developing Countries, but if much of the increase in population will swell the cities, how are the cities to be supplied, except by large power stations?

Secondly, much of the production from renewables' plant will be intermittent. With little progress made so far with large scale energy storage, how can such production cope? That is seen as one of two major question-marks.

The other is availability of land. With more population needing more agricultural land, how certain can one be that there will be adequate land for renewables to produce some 20% of total energy demand by mid-century?

Protagonists are undaunted by such questions, but agree that rapid progress can only be made if governments will it. There are many ways of doing so, from taxation to trading of 'green certificates'. They also argue that prototype projects, being small, are relatively cheap compared to fossil fuel or nuclear plant Furthermore the time span between generations of plant is far lower than for fossil or nuclear plants, so that rapid progress with weeding out unsatisfactory technology becomes possible.

As regards the intermittent nature of power generation, that can be minimised by ensuring a correct mix of intermittent and conventional generating equipment. Profitability calculations do assume lower availability from intermittent plants. The question of land availability could well be serious in some countries, but perhaps in most, more efficient agricultural methods may well release land, even taking the rise in population into account. The availability of new cash crops for biomass may, indeed, spur on better agricultural methods and enrich the countryside.

In summary, it can be said that the case for renewables is stronger than seen by the sceptics, but less certain than believed by the protagonists. It may be ten to fifteen years before the position will be clearer. As will be seen, this is very similar to the nuclear situation, although in this case, there is little antagonism towards renewables and energy efficiency, so that governments have far more freedom to take action and find ways of assisting this industry, should they wish to do so.

Summary.

Arising from the above analysis of the demand and supply situation for energy, how strong is the case for nuclear energy?

On the demand side there are vast uncertainties which cannot be forecast away. As military planners have found out long ago, the reaction to uncertainty is flexibility and that must be an important aim for the supply side. It must be sufficiently flexible to be able to adjust to widely different needs, as the trends into future years become clearer.

Without the issue of climate change, the position would be simple. There are large resources of fossil fuels, especially coal, and the market should be able to decide which fuel has the advantage. That will change from country to country and even region to region. The problems of pollution and energy security will come into such decisions, but how they are dealt with is a political matter which may be resolved differently by different countries. If renewables 'make the grade', they will expand, if not, they will not; the same implies for nuclear energy.

The problem of the greenhouse effect and the international agreements which requires limitations of CO_2 emission has changed this position. To achieve lower CO_2 emission with increasing energy demand, requires limitations of the use of fossil fuels. Admittedly, there is the possibility that ways will be found of sequestering (i.e. storing out of harm's way) the CO_2, but, *as seen today,* this may only make a small difference to the need to reduce the use of fossil fuels.

The only two fuel sources which can substitute for the reduction in the use of fossil fuels are nuclear energy and renewables*. Should the optimistic view about renewables become reality, the case for a large expansion of nuclear energy becomes weak, unless it becomes economically and politically more attractive. However, should renewables develop far more slowly and hit some of the problems, foreseen by the sceptics, there could be a strong case for a nuclear expansion, though, as will be discussed in the next section, only if this became politically acceptable.

In brief, with today's knowledge, we cannot be certain that nuclear energy will definitely be required, nor can we tell that it will never be needed. Thus neither the pro- nor the anti-nuclear position seem tenable. As mentioned in the Introduction, this implies that *both* options should be kept open, though, as will be argued in the Conclusions to this paper, not for an indefinite period.

KEEPING THE NUCLEAR OPTION OPEN.

What does that imply? This section will discuss that question by first considering the public perception of nuclear power and the impact of the changing government policies towards the electricity market. It will then look at what would be expected of the industry and whether and under what conditions, that could be delivered.

The Public, Governments and Politics.

The fear of and opposition to nuclear energy by vociferous sections of the public in many countries is well known. Whether, as the industry strongly believes, such fears are irrational or not is irrelevant. What matters is not 'facts', but perception.

The public, spurred on by environmental groups, is largely concerned by the possibility of the emission of radioactivity and the dangers of the civil nuclear programme assisting the spread of nuclear weapons to countries and terrorist groups. More specifically, the disagreements between environmentalists and the industry concentrate on the following areas:

proliferation of nuclear weapons
safety of nuclear reactors
disposition of nuclear waste

---

* Renewables, such as biomass do emit carbon dioxide, but as growth of the original tree or plant leads to the absorption of CO_2, the net effect is close to 0.

unattractive economics compared to other forms of energy
lack of strong international control of setting safety and security standards and monitoring these.

As mentioned in the Introduction, there is little constructive debate between the opposing parties because both believe that their opponents' views are either illogical or deliberately false. One major problem in achieving a better understanding of each others' position are the differences in the perception of risk. The industry considers the assessment of risk to be a major probabilistic engineering study which arrives at probabilities of major accidents, e.g. that the probability of a major release of radioactivity from reactor of latest design is one in a million reactor operating years. The public tends to be unimpressed, partly because most of us do not readily comprehend statistical methods, but largely because it sees 'risk' in quite a different manner[x]. People ask themselves the question whether there is a trade-off between the possible risk 'to me and my family' and any benefits 'we may receive'.

When it comes to nuclear energy, people see a risk (rightly or wrongly), but, unless connected to the industry, they find it hard to see any benefits. For renewables, on the other hand, many believe they can see the benefits, but hardly any risks (unless they happen to live near a wind farm). So far, the industry has fought back by arguing that the public is being misled and does not understand the true situation, but the result is that it is seen as arrogant - 'we are the experts and we know best' - at a time when there is ever-greater cynicism about experts.

On proliferation, the argument is largely whether or not availability of nuclear power and the research facilities, needed to service the industry, will assist a country's aims to develop nuclear weapons capability, seeing that spent reactor fuel contains Plutonium and that there is presently no ready way of permanently disposing of spent fuel. The recent crises caused by fears that N. Korea may be planning to make use of Pu from spent fuel, just for that purpose, indicate that this is not a barren argument.

The industry suffers from another problem, - secretiveness. It grew out the nuclear weapons industry where secrecy was/is paramount. Although in most countries there is, by now, no connection with the military side, the tradition of secrecy has lingered on, so increasing mistrust by the public.

Such public attitudes are having a considerable influence on governments, at least in democracies. Politicians like to be liked; when it comes to defending State assistance to nuclear power or to renewables, it is quite clear which most would choose.

Lastly, there is the dilemma that Governments wish to make the electricity market sufficiently competitive to be able to leave major decisions to private companies and the free market, whilst they have also committed themselves to improving environmental standards by reducing CO_2 emission and pollution in general. Under conditions of a strongly competitive market, private companies have to take a more short term view than government decision makers, with the result that new capacity would be chosen by narrow profitability criteria. To achieve Governmental environmental targets is likely to need fiscal and/or regulatory measures, which, if not adequately thought through could easily demolish the aims of having a free electricity market. Giving assistance via such measures to renewables could be attacked by users of other fuels, but it may well be acceptable or even popular with the electorate. Doing the same for nuclear energy may, under present circumstances, be unpopular and it would need a brave government to propose that.

In brief, should 'keeping options open' for both renewables and nuclear energy require assistance and good will from governments, the latter would be under a grave disadvantage, unless the nuclear industry manages to change its image with the public at least from 'unacceptable' to 'neutral'. Can that be achieved?

The Nuclear Industry of the Future.

'Keeping the option open' has to imply that, if required, the industry must be able to supply a significant proportion of energy. As an example, the size of the nuclear generating industry for the six WEC/IIASA scenarios are shown in Table 4. below.[xi]

Table 4. Nuclear Capacity under different scenarios. GWe

	Scenarios					
Year	**A1**	**A2**	**A3**	**B**	**C1**	**C2**
2020	**646**	**417**	**732**	**645**	**480**	**605**
2050	**1875**	**782**	**1860**	**1915**	**380**	**1240**
2100	**6415**	**3680**	**6725**	**5700**	**0**	**2750**

1990 capacity - 357 GWe

It will be seen that for the high cases (A1, A3, B and C2), capacity in mid-century would be between four to six time larger than in 1990 with further very substantial growth for the rest of the century. For all these cases the assumption is made that the industry will have resolved the proliferation, waste and safety concerns. But, there must be doubts whether present technology and the present U/Pu fuel cycle is adequate for such expansion.

Use of Present Technology. Large expansion would imply many more reactors and the question is bound to be asked whether even today's designs are safe enough, especially under conditions of mal-operation and lack of adequate maintenance. If such questions cause additional safety provisions to present designs, that would lead to increased capital and operating costs and so to a worse competitive position of nuclear power[xii].

The present resource base for Uranium (Table 2) is quite inadequate for such a major expansion if it operates on a once through basis. However, if experience with oil and natural gas is a guide, far more deposits may well be found if there is an economic incentive to do so. One would then be faced with large quantities of spent fuel which being High Level Waste (HLW) has to be securely stored for decades before being disposed of in deep underground deposits. So far, no such depository is ready to receive this material, largely because the problem of finding geological formations, with a good chance of being stable and impermeable for at least 10,000 years, is rather daunting. The costs are equally daunting with the U.S. Yucca Mountain project scheduled to cost some $30 billion[xiii]. In addition, environmentalists are violently opposed to such storage, believing that they may well turn out to be the Plutonium mines of the future. Yet, under the scenarios assuming rapid growth of nuclear energy, the World may need one new depository every two years.

Should recycle with fast breeders and reprocessing be the choice, the World may have to have some twenty to thirty reprocessing plants if the capacity of plants were similar to the latest plants in France and UK. Such plants are complex and require a very high standard of operation and maintenance; in principle, that should be achievable everywhere, but what safeguards can one devise in the case of corruption, strikes or revolution?

There would also have to be large number of transports taking spent fuel to reprocessing sites and transport back fresh fuel, all of which would contain Pu. However strict the regulations, the chances of negligence somewhere along the complex chain must be considerable and that, the antagonists will argue, may well lead to proliferation or nuclear terrorism.

Unless the nuclear industry can show convincingly that it can operate safely and securely even under unsatisfactory conditions, there seems little chance that world-wide expansion using today's technology will be either acceptable or desirable.

The Alternatives. In most other energy industries the idea that today's technology, already fifty years old and originally developed for quite different reasons, should still be used throughout the next century would be bizarre. Only in the nuclear industry do such concepts appear acceptable. Admittedly, the cost to develop and commercialise new schemes is high costs and so far, nearly all such initial development and proving was done by governments, mostly funded by military budgets. Yet, the resulting attitude of "Progress by design changes - yes, radically new thinking - no" is surely unacceptable for an industry looking towards a dynamic and expanding future.

Technological developments in laboratories and engineering offices in all the major nuclear states are carrying out work, which if successful, could overcome, or at least lessen, many of the barriers to the acceptability of nuclear energy. Most of the work is concentrated in three areas:

1. Improvement of reactor safety, even under mal-operation and maintenance and of the economics of smaller scale reactors.
2. Transmutation of Actinides, incl. Pu with the aim of reducing the quantity of HLW and its half-life from 100,000+ years to a few hundred and never having Pu in a state where it could be misappropriated for weapons use .
3. Explore the potential of making use of the Thorium fuel cycle either by itself or in combination with the Uranium cycle.

As with all research, there can be no guarantee that it will be successful either technically or commercially, but such work does need to continue if the nuclear option is to be kept open. As with most research on energy matters, there have, over recent years, been severe cuts in R. & D. expenditure both by Governments and by private industry [xiv]. Continuation for such work must, therefore, be at risk, especially if there are political pressures to increase research on renewables and conservation and cut nuclear research.

World-wide, there are a large variety of nuclear research projects; not all can be developed and sooner or later a choice has to be made where to target research funds. For now, there is no mechanism for such choice which involves governments and private industry. Even though there is a great deal of scientific and technical collaboration, it is likely that each country will take its own decisions, which for many countries may be to cease the work because of the high costs of going it alone. If that were to lead to a considerable slow-down or even cessation of much of the nuclear research, the nuclear option will slowly disappear.

Not Keeping the Nuclear Option. Anti-nuclear groups wish to phase out nuclear power as soon as possible, but unless there is much increased pressure owing to a serious accident, that is unlikely to happen in most countries owing to the difficulties and costs of finding replacements. More realistically, a fade-out may happen over the next twenty to thirty years as nuclear plants, which reach the end of their lives, are replaced by non-nuclear plant and there are no robust longer term development plans for nuclear facilities. In other words, the industry would slowly run down, but that would not be its end. It would still have the task of disposing of all the nuclear waste and cleaning up contaminated sites, both from the civil and military programmes. That may take beyond the end of the next century and cost hundreds of billions $ if today's technology were used.

Experience in other fields has shown that research can achieve considerable reductions in cost and this should also be the case here. Some of such research may well be similar to what is being done now. Closing such R. & D. would give a message to future generations that we are leaving a difficult mess behind for them to clean up. Not a good advertisement for our intent to achieve 'sustainability'.

Summary. The key to keeping the nuclear option open is not expansion based on today's technology, but a concentrated effort in the field of R. & D. to determine whether it is possible to develop a nuclear fuel cycle and nuclear plant which would meet most of the concerns of the public. There are indications from laboratories that this could be achieved, but far more work is needed before one could be certain.

There is time to proceed with such work, but, unless the industry is able to achieve greater acceptability by the public, there must be doubt whether the necessary funds will be made available. The need for continuing research on waste disposal, whether the nuclear option is or is not kept, could be an argument for continuation of funding.

CONCLUSIONS.

With the WEC/IIASA scenarios as the basis for this paper, the following conclusions are reached:

1. The rate of growth of demand for energy during the century is highly uncertain.
2. Unless such growth is extremely high, there appear to be adequate resources of fossil fuels for a number of decades; when taking more unconventional sources into account, adequacy may last for the century. If curtailment of greenhouse gas emission is to continue, use of fossil fuels will have to be reduced. There would have to be substitution by faster improvement of energy efficiency (conservation), by greater use of non-fossil fuel.
3. There is disagreement whether conservation and renewables would be able to fill the gap left by the reduction of fossil fuels use, or whether nuclear power would be required as well. At the present time there is inadequate information to reach a conclusion on this and it may well be ten to fifteen years before the real potential for renewables can be established. Owing to this uncertainty, there is a good case for keeping the nuclear option open, at least until more experience has been gained with renewables.
4. There can be little doubt that the unpopularity of nuclear power is a grave obstacle to its future. State assistance for research on and commercialisation of renewables is widespread and popular; assistance for nuclear power could well be politically difficult.
5. Keeping the nuclear option open implies that nuclear technology should be made ready to take a significant proportion - say over 10% by mid-century - of global energy needs. It is doubtful whether today's fuel cycle and technology would be sufficiently acceptable to be allowed to embark on such an expansion. However, laboratory work in a number of countries indicate that solutions to many of the public's concerns could well be found by changing the fuel cycle and making use of different types of reactors.
6. In the light of the industry's unpopularity, there have to be doubts whether there will be adequate funding for such work. More international collaboration and means of choosing which of the many research schemes should be given priority may make whatever funds will become available more productive. The question of the nuclear option will have to be reviewed when more is known about future energy trends and the progress with conservation and renewables. The nuclear industry may have some 10 to 15 years to make a strong and politically acceptable case for its continuation.
7. Should nuclear energy be phased out, the work to dispose of nuclear waste, to decommission plant and to decontaminate the environment may well take a century and be very expensive. Without a profitable industry to back up such work, the burden to carry it out will fall on the next generations.

8. R.& D. to improve such clean-up will be essential and may well be similar to work now being done to find acceptable means of disposing high level waste. That may make the case for continued funding of nuclear R. & D. somewhat stronger.

The future of the industry will be more determined by R. & D. than by orders for plant based on today's technology. Before governments might accept to fund such work, it will be necessary to show that public opinion is at least neutral towards nuclear power. The first step towards achieving this could be to establish a dialogue between the industry and the (more moderate) opposition. The purpose should be

* To reach a better understanding of each others' position
* Reach agreement that owing to the vast uncertainties of the future, the nuclear option should be kept open.
* Find an outline of an acceptable nuclear industry, so that research could be focussed on developing the necessary technology.

Literature Cited:

[i] N. Nakicenovic and J. M. Jefferson . *Global Energy Perspective to 2050 and Beyond.* World Energy Council, 1995

[ii] Interlaboratory Working Group on Energy Efficient and Low Carbon Technologies: "Scenarios of U.S. Carbon Reductions.", Ch.1, Table 1.1,U.S. Department of Energy, 1997

[iii] M.J. Pink, "Exploration and appraisal technology - maximising rewards by integration" Shell Selected Papers 1993.

[iv] P. Kassler, "Energy for Development", p 8, Shell Selected Papers, 1994, Shell International Petroleum Co

[v] C. J. Campbell, "*The Resource Constraint to Oil Production: The Spectre of a Chronic Supply Shortfall*", pp 29 - 59 in "World Energy - A Changing Scene" Editor M Kürsten,. Schweizerbart'sche Verlagsbuch-handlung, Stuttgart, 1996.

[vi] P Tijm et al. "*Liquefied Natural Gas: reducing costs.*" Shell Selected Papers, June 1995

[vii] U.S. Dept. of Energy, http://www.hydrate.org

[viii] Web site http://www.eren.doe.gov./

[ix] Helena L. Chum. "*Two Decades of Progress in R. & D. and Commercialisation of Renewable Energy.*" Proceedings of a Panel Discussion on 'The Enduring Nuclear Fuel Cycle' at the American Nuclear Soc., November 1997, National Technical Information Service, U.S. Dept of Commerce.

[x] H. Newby, "*Risk Analysis and Risk Perception.*" Transaction of the Institution of Chemical Engineers, Part B, Vol. 75, No. B3 (August 1997)

[xi] M. Jefferson "*Nuclear Power in global energy perspectives of the twenty-first century.*" Nuclear Energy, 1997, **36**, No. 2, Apr. pp115-122

[xii] Richard Wilson: 'Over-regulation and other problems of Nuclear Power' presented at the International Conference on Environment and Nuclear Energy, Washington D.C., 29th October 1997.

[xiii] T. Myers et al: "*Accelerator Assisted Repositories*" ISTC International Seminar 'New Approaches to the International Fuel Cycles", Sarov, Russia, June 1998.

[xiv] The President's Committee of Advisers on Science and Technology: "*Federal Energy Research and Development for the Challenges of the 21st Century.*"

NUCLEAR POWER AND SUSTAINABLE DEVELOPMENT

Mr. Stig Sandklef

President Vattenfall Fuel
SE-162 87 STOCKHOLM

INTRODUCTION

Nuclear Power is a new, innovative technology for energy production, seen in the longer historic perspective. Nuclear technology has a large potential for further development and use in new applications. To achieve this potential the industry needs to develop the arguments to convince policy makers and the general public that nuclear power is a real alternative as part of a sustainable energy system.

This paper examines the basic concept of sustainable development and gives a quality review of the most important factors and requirements, which have to be met to quality nuclear power as sustainable.

This paper intends to demonstrate that it is not only in minimising greenhouse gas emissions that nuclear power is a sustainable technology, also with respect to land use, fuel availability waste disposal, recycling and use of limited economic resources arguments can be developed in favour of nuclear power as a long term sustainable technology.

SUSTAINABLE DEVELOPMENT

World population is projected to grow to 10 billion people in 2050. Most of this growth will occur in developing countries where the future generations have a right to improve their quality of life, have enough food, clean water, clean air and an environment which is not polluted to the extent it is today. Releases of greenhouse gases must be reduced even from current levels to avoid climate change. Developing countries will suffer disproportionately from the affects of climate change, as it will cause problems in basic areas such as their ability to grow enough food to take care of their growing population.

Agenda 21, adopted at the Rio Conference on the Environment and Development in 1992 notes that "energy is essential to economic and social development and improved quality of life.etc. All energy sources will be needed to use in ways that respect the atmosphere, human health and the environment."

Preparing the Ground for Renewal of Nuclear Power, edited by
Kursunoglu *et al.* Kluwer Academic / Plenum Publishers, New York 1999.

We all realise that these prospects pose enormous challenges on sustainable development and particularly on the systems for energy production, distribution and use that will be needed in the future. We need to examine the requirements for future energy systems in this context and within the concept of sustainable development. In 1987 the Brundtland Commission gave the generally accepted definition of sustainable development as follows:

"Development that meets the needs and aspirations of the present generation without compromising the ability of future generations to meet their own needs"

When measuring sustainability for energy production against this principle the following factors should be considered:

- Fuel availability and use
- Land use
- Environmental effects of waste disposal
- Recycling possibilities
- Availability and competitiveness, including external and social costs
- Climate change

Let us examine these factors applied to nuclear energy:

FUEL AVAILABILITY

Uranium is the basis for nuclear fuel. There are abundant reserves available and they have no other constructive use. The natural biological cycles rely on hydrogen, oxygen, carbon and nitrogen. The use of uranium does not interfere with any of these natural cycles and therefore leaves these valuable resources for other uses. Today estimated reserves of uranium equal about 250 years of consumption, with recycling of uranium and plutonium this figure can be extended to 10,000 years.

There are known and estimated additional reserves of uranium which would extend this figure even further allowing for significant increases in nuclear programmes and still maintain, what is in practice an unlimited fuel resource base.

Table 1. Fuel Availability. Estimated Uranium reserves

Category	Mio tU	Years of present consumption
Stocks	0.2	3
Weapons materials	0.6	9
Known in ground	4.5	70
Estimated in ground	11	170
Estimated in phosphates	22	340
Sea water	4 000	60 000

We can conclude from these figures that uranium availability does not restrict the use of nuclear power now or in the longer term even with significant increases of nuclear programs. Even if costs for uranium would increase by a factor of 10 it would still be less expensive than any of the competing fuels.

Uranium is found in abundance in a number of politically stable countries such as Canada and Australia. Trade with uranium producers is very stable, and most European utilities have long term contracts with very little price fluctuation, and maintain inventories of one to three years worth of nuclear fuel, preserving security of supply.

The energy intensity of uranium is beneficial from an environmental point of view:

* 1kg of firewood produces about 1 kWh(e).
* 1kg of coal produces 3 kWh(e)
* 1kg of oil produces 4 kWh(e)
* 1kg of uranium produces about 50,000 kWh(e)

The volume of nuclear fuel, its transport, land use, storage and quantity of waste, are all much smaller than for fossil fuels.

One tonne of uranium produces as much energy as 17.000 tonnes of coal. CO_2 equivalent emissions for the entire nuclear fuel cycle - from uranium mining to waste disposal - range from 10-50 g/kWh which is about the same as for wind power. Comparable factors for the fossil fuel cycle range from 450 to 1200 g/kWh.

Nuclear power plants can make a future contribution to sustainable development by burning nuclear weapons material to generate electricity. The incorporation of weapons material into nuclear fuel in civil reactors will reduce the risks of proliferation and global security is improved. The material would no longer have weapons potential, because in the commercial fuel cycle it is mixed with uranium and surveyed by IAEA and in Europe by EURATOM safe-guards.

The costs of nuclear fuel for an operating nuclear power plant are only about 8 to 12 percent of total generating costs, a minor expense when compared to the 60 percent of the cost of natural gas in gas fired power plants.

LAND USE

Regarding land use, for comparison purposes a 1000 MW plant was selected. A nuclear site requires 1-4 km^2, a solar or photovoltaic park requires 20-50 km^2, a wind field 50-150 km^2 and for biomass 4000-6000 km^2. Even then a solar or wind park will only have availability factors ranging between 20-40% while Swedish nuclear plants are averaging around 78% in 1997.

Apart from the fact that solar and wind can only be used in places where conditions are favourable and consequently require investments in back up production capacity, the enormous land requirements of biomass make its use in a large scale possible only in countries sparsely populated and with favourable climatic conditions. Many developing countries simply can not afford to set aside large areas for biomass production when the highest priority for land use is food production.

There are projections that in 2050 half of the world's population will live in large cities. This emphases the need for concentrated energy production systems and in practice the large scale use of wind, solar and biomass for the energy needs of these large cities are ruled out. Large-scale use of biomass will be restricted to countries where climate and low population density allow the extensive land requirements (Sweden, Canada, Russia and a few more). The fact that nuclear energy requires a minimum of land will be a strong factor in favour of nuclear power in the future.

ENVIRONMENTAL EFFECTS OF WASTE DISPOSAL

A comparison of the amounts of waste generated by the operation of 1000 MW-nuclear and coal fired power plants results in the following figures:

7.000 tons of coals are need per day or about five trains loads. The nuclear plant will use about 80 kg of uranium per day. Even with sophisticated clean coal technology the coal plant will emit per year about

* 900 tons of SO_2
* 4500 tons of NO_X
* 6.5 million ton of CO_2
* 400 tons of various heavy metals including arsenic and mercury.

The nuclear plant will produce about 20 tons of spent fuel per year. The industry always works to contain any radiation, whether in the fuel rod, in the reactor during operation, or in an engineered barrier system for waste.

If, throughout the lifetime of an average citizen of an OECD country, all of the electricity he or she used were to be generated by nuclear fuel, the vitrified high level waste produced is about the size of a kiwi fruit. Radioactivity decays over time, toxic materials like heavy metals last forever.

Waste is generally managed in two ways, it is either contained or dispersed. Because of the large quantities of fossil fuel waste, such as greenhouse gases, most of this waste is dispersed directly into the atmosphere because there is no practical alternative. Nuclear waste quantities are sufficiently small that our industry can to manage its waste through containment. The principle-disposal options are in surface-based engineered facilities or in previously mined areas or in deep geological repositories.

In many countries, the industry needs to convince the public and politicians that major progress in handling, treatment and storage of nuclear waste has been made in recent years. It is the view of the European nuclear industry that there are no outstanding technical issues to be solved in the industrial management of radioactive waste. European industry has the technological lead in this area, as both reprocessing and direct disposal methods have been demonstrated.

Economically our waste management programmes should give nuclear a competitive advantage over other forms of electricity generation. The nuclear industry's waste has always been accounted for and taken care of. It has never just been released into the air.

An OECD/NEA study which as assessed the costs of waste management concluded that costs ranged from 0.4 to 1.8 mills per kWh compared with total nuclear generation costs for new plants, in the same countries, ranging from 28 to 54 mills per kWh. Therefore we can conclude that waste generation is both minimal and not a determining factor in total generation costs.

In the UK which was the first country in Europe, using nuclear power, all the radioactive waste produced from all sources by 2030 will be equivalent in volume to about three weeks' worth of British garbage. 90 percent of this nuclear-related waste is low-level waste material, which is only slightly contaminated.

Financial reserves are accumulated for decommissioning and waste storage facilities, through electricity prices in most European countries, as part of the provisions of the industry to finance both the handling and treatment of waste as well as the later decommissioning and dismantling of the nuclear facilities. This practice is based on another criteria of sustainability: that the people who have benefited from the electricity should also be responsible for the waste management.

Technically the industry has demonstrated its ability to condition all forms of nuclear waste for storage and to complete geological repositories needed for the final disposal of nuclear waste.

RECYCLING POSSIBILITIES

Industry recycling of its spent fuel contributes to waste reduction. Even though spent fuel from power plants can be safely disposed of, as high level waste, it contains large amounts of energy. 97 percent of the total volume may be recycled, thereby replacing the need for a corresponding new production of uranium fuel. By recycling uranium and plutonium as reactor fuel up to 60 times as much of the energy content can be utilised in fast neutron reactors. This technology is available and already used on a large scale in some countries. The industry is also committed to reducing waste volumes by compacting, drying and burning in incinerators.

AVAILABILITY

Nuclear technology is not only available today and demonstrating its environmental and cost benefits in many countries, there is also a dynamic development of evolutionary and new reactor concepts in several countries.

There are 428 reactors, representing 344 774 MWe in operation in the world today and 30 reactors under construction, making nuclear power the third largest electricity generation technology in the world.

New reactor designs have been or are being developed to meet the needs of future nuclear power expansion. The EPR (European Pressurised Reactor) and similar developments for a European Boiling Water Reactor, the ABWR (Advanced Boiling Water Reactor) in Japan and the US and CE 80+ (an advanced pressurised water reactor) in Korea and the US may serve as examples of a dynamic development.

COMPETITIVENESS

A key factor to achieve sustainable development is the economic use of available resources including financial resources.

There are risks associated with any industrial activity. There are also risks associated with a lack of sufficient energy production sources. In the poorest countries 25.000 children die every day because of a lack of clean water, a problem which could be remedied with pumps, purification facilities and the energy to operate them. There have been a number of studies of the comparative risks of electricity production. A recent example by the OECD/NEA demonstrates that the latent risks of disease or accident from coal based electricity generation is 100 times more dangerous than the operation of a light water reactor in normal conditions.

The full external and social costs of any energy production technology should be taken into account, thereby including environmental effects for the entire fuel chain and occupational as well as public effects on a local, regional and global scale. Extensive studies are carried out to quantify these external costs and even if results are still uncertain the assessments can highlight the basic characteristics of different energy production technologies. The comprehensive ExternE project, carried out by the European Commission in collaboration with the US Department of Energy examined externalities for complete energy chains and the results are shown in the following table.

Table 2. Total costs of electricity production in cECU/kWh

Technology	External costs	Operating, fuel, financial costs	Total
Coal	2.0	5.0	7
Oil	1.6	4.5	6
Gas	0.36	3.5	3.9
Wind	0.22	6.0	6.2
Hydro	0.22	4.5	4.7
Nuclear	0.04	3.5	3.5

The operating and financial costs of the different technologies vary in different countries depending on local conditions, assumptions on interest rates etc.

The external costs of nuclear power cover the potential costs of large accidents taking into account the low probability of such accidents.

In general hydro and natural gas with modern technology and nuclear are very close for most scenarios counting only operating and financial costs. Taking into account the external costs given above natural gas and nuclear are the most favourable options.

Estimates of external cost of CO_2 emissions (climate change effect) are very uncertain and range from 10 to 25 USD/ton coal. Assuming 15 USD/ton coal gives an external cost of 0.5 cECU/kWh for coal and 0.3 cECU/kWh for natural gas (included in table 2). The upper value would increase these figures with more than half. This leaves nuclear power as the most economic alternative when all costs are taken into consideration.

CLIMATE CHANGE

Up to 75% of the CO2 emissions from human sources in the atmosphere are caused by the burning of fossil fuels; coal, oil, and gas and the main remaining component is the deforestation in the large rain forests.

As stated above, the use of nuclear energy results in a release of CO2 equivalents comparable to those of wind power and orders of magnitude less than the burning of fossil fuel.

Nuclear energy and hydropower are the only major sources of essentially carbon-free, economic electricity generation, which could be expanded today at a significant scale.

While our scientific understanding of global warming continues to grow, it is important to rely increasingly on those sources of electricity which do not contribute any greenhouse gases to the atmosphere, like renewables and nuclear energy. In the European Union and other developed areas of the world, continued reliance on electricity must be a key component of any sustainable development policy. A balanced energy policy, involving the reduction of fossil fuel use, a focus on energy conservation and efficiency improvements, and an increasing reliance on non-greenhouse gas emitting energy sources, will lead to fewer environmental hazards, while maintaining our standard of living.

ENERGY SCENARIOS

Any projection for long term energy use must take into consideration the large unbalance between industrialised and developing countries of today and the strong driving forces, both in terms of population growth and demand for improved living conditions. Any projection for the longer-term future will predict significantly higher figures for energy consumption.

The World Energy Council/IIASA has recently published scenarios, which are summarised in table 3.

Table 3. Energy Scenarios until 2050

	Unit	Today	2050
World population	billions	5.7	10
World primary energy	Gtoe	8.4	20-25
Share of fossil fuels	%	90	59-73
World electricity gen.	Twh	13 000	31 000-42 000
Share of nuclear	%	17	11-38
Carbon dioxide Emissions	GT	24	34-55

The scenario with the lowest emission rate for greenhouse gases (34 GT compared with to-days 24) does not meet any of the objectives set in Kyoto. In the high nuclear scenario (38% of 42000 Twh) there is a growth of a factor of 6 while the low nuclear scenario practically maintains to-days capacity. An increase of nuclear power capacity by a factor of 6 is not in itself enough to reduce climate change. But it must be clear to everyone that a significant increase in the use of nuclear energy is a vital part of any long term energy policy. And it can be done. The industry has the capacity to expand.

CONCLUSIONS

I have demonstrated that nuclear power is in all aspects a sustainable technology, which could serve us in the long term with minimal environmental effects and at minimum costs to the society. And the challenge can be met.

But to achieve it we need political leadership, to support and develop the institutional and legal framework that is the basis for a stable and long-term energy policy. And we need industry leaders to stand up for nuclear power, to create a new industry culture of openness and communication with the public that is necessary to get the public acceptance that we have failed to do so far. The basic facts are all in favour of nuclear power. We should use them!

I want to acknowledge the valuable help from Linda Gunter, FORATOM and Carl-Erik Wikdahl, Swedish Atomic Forum in preparing this paper.

PRESENTATIONS OF THE INTERNATIONAL CONFERENCE ON PREPARING THE GROUND FOR RENEWAL OF NUCLEAR POWER

Robert A. Krakowski

Systems Engineering and Integration Group
Technology and Safety Assessment Division
Los Alamos National Laboratory
Los Alamos, New Mexico 87545, USA

INTRODUCTION

Background

This 21st in a series of energy conferences since 1974 focused on the following questions:

- Will nuclear power plants (NPPs) be competitive at the time of replacement (next 10-20 years) with other energy sources?
- Will public acceptance of nuclear power be sufficiently improved to allow the construction and operation of otherwise economically competitive NPPs?
- Will the positive environmental attributes of nuclear energy with regard to reduced greenhouse gas (GHG) emissions and increased energy security play an important role in future policies promulgated by decision makers and the public?

To address these questions, 111 delegates gathered for this two-day conference to listen to twenty-eight 20-30 minute presentations dealing with four key issues embodied in these three focus questions: economic competitiveness (eight presentations on economics, politics, and evolution of NPPs); public acceptance (seven presentations); resources (four presentations); and environment (four presentations). Combining with five opening presentations (E. Adamov and G. Charpak were not in attendance) and six panelists' presentations leads to a total of 24 formal presentations on the subject of NPP economic competitiveness, public acceptance, resource durability, and environmental effects. The national/organization breakdown of these formal presentations is as follows: EU(1), France(6), Germany(2), IAEA(2), Japan(1), OECD(1), Sweden(2), United Kingdom(1), United States(9), and WEC(1) - a total of 26 formal presentations (including the concluding panel remarks) from six countries and four international organizations were delivered.

Preparing the Ground for Renewal of Nuclear Power, edited by
Kursunoglu *et al.* Kluwer Academic / Plenum Publishers, New York 1999.

Conference Overview

Generally, most papers described the *status quo* of nuclear energy (NE) at a relatively high (qualitative) level and strongly supported the positions of the NE industries or government organizations sponsoring the respective speakers. While no formal attempt was made to generate a conference consensus, the following key positions emerged from both papers presented and ensuing discussions.

- Economic competitiveness is the premier issue for NE; without NPPs that are competitive with 0-20 year alternatives (*i.e.*, CCGTs), issues of public acceptance, environmental attractiveness (very low carbon intensity, *e.g.*, kgC/kWeh), and energy security become irrelevant.
- Nuclear energy cannot rely on the attribute of very low carbon intensity (*e.g.*, kgC/kWeh) compared to fossil energy to survive and grow beyond the 10-20 year phase-out period envisaged for existing NPPs.
- To some degree certain short-term positions suggested for NE are not consistent with certain longer-term claims; the use of conventional or evolutionary LWR (ELWR) designs based either on once-through or plutonium recycle are inconsistent with both fuel and repository demands under any scenario where NE makes a significant contribution to the mitigation of greenhouse gas emissions.
- The public-acceptance issue permeates all NE decision making in all NE and non-NE countries, with only economic prospects of an NE phase-out in those countries that are heavily dependent on this technology (*i.e.*, Sweden, France, Germany, *etc.*) tempering reactions associated with anti-nuclear public-acceptance forces.
- The NE phase-out posture of the new Social Democrat government of Gerhardt Schroeder in Germany received numerous questions, to which the reply was: a) strong support exists for NE in general and the Franco-German EPR project, in particular, on the part of French Prime Minister; b) contracts are contracts (with respect to the EPR project); c) a phase out, if it happens, would occur on a time scale that is (probably) much longer than that of most governments; d) good technical work must proceed as far as possible without undue political hindrances or course changes.
- Proliferation risk associated with the civil fuel cycle is primarily a U.S. issue and is not part of the set of "cardinal issues" being addressed by NE programs in other OECD countries (*e.g.*, economic competitiveness, waste, and safety).
- Operating, maintenance, and fuel-cycle costs have decreased dramatically for NE, and ~30-40% reductions in capital costs are needed for NE to become strongly competitive with present (low-fuel-cost) fossil energy; advances in computer applications (*e.g.*, "smart" components, design, installation), application of Probabilistic Risk Analysis as a design tool for simplification and the development of risk-based performance-based regulatory processes may reduce capital costs; the need for a renaissance in government-industry relationships, as well as the realities of NPP cost centers and the connectivity between safety and reliability, however, mitigate against much relief from these approaches to capital-cost reduction.
- A general call was made for small NPPs that could compete on a distributed electrical power grid expected to evolve in the developing countries under competition from systems like CCGTs; this call seemed to ignore economies-of-scale drivers that has led to plants of EPR sizes (1.5 GWe), as well as technological innovations in electricity distributions and networks that might pull away from this trend towards distributed generation/network systems.

- Aside from difficult-to-quantify beyond-direct-cost (BDC) issues, and until these issues are better and more uniformly internalized, NE must control capital costs and project competitiveness in a market that generally places little value on the far future. Meanwhile, policy issues must be more clearly and quantitatively incorporated into the decision-making process to include security of energy supply, social concerns (as reflected in negative BDC impacts related to adverse public acceptance of NE technologies), and long-term environmental impacts related to GCC and both water and land use.
- Proliferation is not a major issue in Europe, and the NE community should move on to other issues; proliferation is or is not an issue in the U.S., depending on whom is expounding. Generally, however, a consensus probably could be found that proliferation is primarily a U.S. problem.
- R&D on the effects of low levels of radiation and on the application of transmutation technologies are two areas of strong payoff and should be pursued vigorously; with respect to the latter, transmutation technologies can have strong roles under either NE phase-in or NE phase-out scenarios, and such technologies can form a valuable technical bridge between advocates of either vision of the future.

CONFERENCE DETAILS

Opening Remarks

The present situation for nuclear energy (NE) was vividly and accurately captured by Jean Couture when he described a process of building a road to the future while simultaneously using it to move forward. Behram Kursunoglu noted that the present meeting was the fourth to be held in Paris and called for the use of NE to combat, mitigate, or avoid serious environmental effects associated with the excessive use of fossil energy (FE), as well as continuing strong actions that stem the "second phase of the cold war" related to the acquisition of nuclear weapons by rogue nations.

Chauncey Starr (EPRI, Director Emeritus) presented through video tape the need for a rational energy policy based on a balance between the realities of global population growth, pollution constraints, and the drive for global prosperity; each of these three pillars must support a balanced set of policies that are workable and stable. Nuclear energy must play a major role in that policy, and NE must develop a strong counter campaign against the anti-nuclear movement that has successfully intervened to interfere with such a role. A call was made for a comparative benefit-to-cost analysis for all energy sources, and this assessment must be organized now. This attack on a highly motivated anti-technology mood and movement must be launched through the schools to develop a more balanced understanding and appreciation of the connectivity of technology, radiation, and energy. These points were punctuated with a quantitative needs assessment out to the year 2100 based on a doubling of the world population by people who at a minimum desire a *per-capita* energy utilization that increases by a factor of (at least) 2.5. The factor of ~5 increase in total primary energy (TPE) consumption that results raises critical questions of energy supply, cost, and environmental impact. The visionary goal of supplying these needs through renewable energy (RE) sources is central to the anti-nuclear environmental movement, which was characterized as an "anti-social doctrine that can only create havoc to global welfare if seriously implemented." Public trust in the integrity and competence of the nuclear industry, coupled with a reasonable expectation on the part of the public that NE is an important part of a balanced choice, is crucial. This call for increased public acceptance set the tone for a key issue faced by this conference.

L. Neel called for NE to be put back into perspective, particularly with respect to the effects of low levels of radiation on humans. Since neither science nor public opinion has clear evidence as to the biological impact of low radiation doses, a stronger R&D campaign must be mounted to resolve this crucial issue that strongly impacts public acceptance. It was noted the recent campaigns on automobile safety began only when measurements and experiences were accumulated, and a similar process is expected for low-level radiation effects.

M. Doucet (Secretary General WEC) claimed that the NE industry has been its own worst enemy through insularity and not being sufficiently familiar with members of the family that comprise competitive energy mixes. In reporting on the September 1998 WEC Houston Conference, the key issues for NE in descending order of importance are economics, public acceptance, and (then) global climate change (GCC); "if NEs rationale is based only on CO_2, NE will lose," in that other energy sectors are advancing strongly on the CO_2 issue more than NE is advancing on its own problems. In Canada (Doucet is Canadian), the NE industry is also its own worst enemy through generating dual rationales for heavy-water plants and related government subsidies; recently, the Canadian Prime Minister called for the shutting down of all of Canada's NPPs because of cost overruns, lack of managerial transparency, systems problems, and crisis management. Even with South Korea doubling its NPPs (14 $\rightarrow$ 30) and Japan adding 55 new plants, the worldwide share of NE is decreasing, and this decrease is occurring in spite of a 50% increase in primary-energy (PE) demand by 2020. The economic viability of NE depends crucially on discount rate, with the recent NEA/OECD study on projected costs of generating electricity in 2005-2010 indicating that in ten out of twelve cases NPPs were competitive for a 5%/yr discount rate, but this number was reduced to zero for a 10%/yr rate. The major WEC energy studies indicate 90% growth in PE demand occurring in developing countries, with a strong need to examine niche markets in these demand-growth area, where distributed, less-organized, and more-localized generation will be needed. In all WEC scenarios under these conditions, smaller and smarter NPPs were claimed to be needed. Compared to the 1992 WEC study, the reports from Houston indicate strong advances in clean coal technologies, strong in-roads being made for natural gas, and the prospects of ~$16/bbl oil from unconventional (tar-sands) oil. Furthermore, strong growth in the transportation sector is projected for the developing countries, and oil is redefining itself in the context of this shifting transportation sector. It was charged that NE is not sufficiently aware of these changing markets. Growth in desalination and cooling demands in the Middle Eastern (ME) markets were identified as opportunities for NE. Furthermore, demand growth in this area could cause the ME to cease being a net exporter, and resulting shifts in oil markets will set the real price of oil, and whether NE is competitive. Lastly, the whole idea of an "NE Industry" was questioned, in that most industries are both multi-national and multi-energy. The competitiveness of NE must be married with the global CO_2 issue in the form of smaller plants to enter the carbon-trading business by purchasing all carbon permits, particularly those generated by the "Russian CO_2 bubble"; the resulting revenues would be used to support small-reactor development and deployment. This somewhat controversial talk (relative to most of the others in this conference) suggested that: a) NE has not moved with the energy market; b) the reason why NE has not decreased costs where all forms of FE and RE have must be understood; and c) while government interventions into energy markets and related short-term views may be necessary to assure a balanced mix, NE can rely no more on government intervention for its survival.

THE MEETINGS

SESSION I

NUCLEAR COMPETITIVENESS - THE ECONOMICS AND POLITICS OF NUCLEAR ENERGY

Moderators: S. T. Brewer, President, S. Brewer Enterprises, Inc. Connecticut.
J. D. Levi, Senior Vice-President, Framatome.
Annotator: J. Paffenbarger, Electricity Expert, International Energy Agency.

Dissertators: S. T. Brewer, President, S. Brewer Enterprises, Inc. Connecticut.

"Nuclear Business and Finance in the Context of Massive Restructuring of the Electric Utility Sector in the U.S.

Within the U.S., it was claimed to be of far greater importance to deal with market forces than with the technology; NE today does not need another new reactor. This short-term view is consistent with that of a free market, wherein the present drive for deregulation represents a real "sea change". The history of electric utility regulation, deregulation, restructuring, *etc.* in the U.S. was reviewed: a) 1920 FPA (sorted out roles of generator and government); b) 1935 PUCHA; c) 1978 PURPA (competitive rate for non-utility generator); d) 1992 EPA (exempted wholesale generator within utility generator); and e) 1998 States' legislations enabling power to be sold at the retail level. The above-noted "sea changes" are characterized by: a) mandatory rate reductions, b) unbundled utility functions, c) auctions of generating assets, d) arisings of uncompetitive ("stranded") assets, e) creation of the "standard offer", and f) special subsidies for RE sources. The impact of these changes on NE in the U.S. is reflected in zero new construction, rapid rates of NPP ownership changes (~40 in the next 5 years), and incompleted plants because of marginal economics. This situation has created enormous investment opportunities for entrepreneurs who are able to purchase existing NPPs for ~10% of the original investment cost. For example, purchase of a 1-GWe plant for $100M (original cost was ~$2B), operating that plant efficiently (85% capacity) for 15 years, contributing to a ~$500M D&D sinking fund, allowing for one steam-generator replacement, paying nominal taxes, and selling power at the mandated "standard-offer" price will yield a $1.1B profit in 15 years of the (assumed) remaining plant life. A break-even purchase price for this "spreadsheet" financial experiment is ~$500M. Who is paying for these enormous profits? - ultimately the individual rate payer. The state regulatory body (PUC) will eventually establish a floor for these auction prices that minimizes the stranded costs paid by the utilities and rate payers.

P. Savelli, Deputy Director General, NEA

"Nuclear Competitiveness Worldwide"

Following a theme that was repeated many times at this conference, economic competitiveness remains the key factor for the decision maker in offering specific energy mixes to the market. The rising importance of sustainable development, however, brings non-economic elements into this decision process. The focus of this presentation incorporated both aspects: a) projected costs for generating electricity; and b) issues and constraints that go beyond direct cost (BDC, *e.g.*, accident risk, security of supply, *etc.*). The recently issued NEA study entitled "Projected Costs of Generating Electricity in 2005-2010: Increasing Competitiveness of Gas, Versus Coal and Nuclear," [NEA/COM(98)8,

October 9, 1998] provided a basis that uses a common methodology (levelized lifetime costs) to analyzed economic and performance data provided by both (OECD) member and non-member countries. Specifically, 15 OECD countries (Belgium, Canada, Denmark, Finland, France, Hungary, Italy, Japan, the Republic of Korea, the Netherlands, Portugal, Spain, Turkey, the United Kingdom, and the United States) and 5 non-OECD countries (Brazil, China, India, Romania, and Russia) were queried on the cost of electricity generation from nuclear, coal, gas, biomass, solar (photovoltaic), and wind. Twelve countries provided NE costs, which varied widely because of design, siting, material, and labor costs; standardizing assumptions included 40-year plant life, 75% capacity factors (conservative, in that values in excess of 80% are achieved and plant lives of 60 years are targeted by newer NPPs), and two values of discount rate (5 and 10%/yr). For the 5%/yr discount rate, all NPPs were cheaper than FE plants for all but two of the 12 countries, but at 10%/yr these capital-intensive NPPs did not compete in any of the countries examined. Fuel price remains a strong suit for NE: <25% COE for NE, compared to 75% for natural gas and 50% for coal. Over the past five years, the decrease in levelized cost for NE has been in the range 2-27%, compared to 6-34% for coal and 16-54% for natural gas. Costs BDC or external to those usually borne by the generator must be more thoroughly incorporated into these kinds of market assessments; these externalities include macroeconomic impacts (employment, price stability, short- and long-term price elasticities, balance of payments), strategic factors, and the environment. The costs of accidents and waste disposal must also be included, and many of these external costs are internalized by NE and not (yet) by FE, which could significantly shift the relative advantages of NE and FE, even for high discount rates. Costwise, NE can tolerate only a few major accidents, and the human-health risks, while difficult to compare, are much less than those for hydro-electric or coal-electric. The BDC impacts related to energy security are equally difficult to assess, but an increase of NE market share to 30% would increase energy costs by only 4%, while reducing significantly the economic risks of future price increases. Aside from these difficult-to-quantify BDC issues, and until these issues are better and more uniformly internalized, NE must control capital costs and project competitiveness in a market that generally places little value on the far future. Meanwhile, policy issues must be more clearly and quantitatively incorporated into the decision-making process to include security of energy supply, social concerns (as reflected in negative BDC impacts related to adverse public acceptance of NE technologies), and long-term environmental impacts related to GCC and both water and land use.

H. Maeda, Executive Managing Director, Kansai Electric Power Co.

"Future Competitiveness on Nuclear Electricity in Japan"

The theme of reduced NE costs was again replayed in the contexts of both total costs and fuel-cycle (FC) costs - a most important goal. At the Kansai Electric Power Company (37 GWe, 138 TWeh/yr in 1997, 11 PWR units), COE has decreased by 20% over the last three years, and these decreases have been driven primarily by the 70-85% increase in plant availability. Some of this increase results from operational policies that favor the generation of more NE electricity than FE electricity (claimed to be only 25% for FE plants). The 80-85% availability limits presently are determined by an operating period between maintenance outages (13 months) that is set by regulations. Staff morale, steam-generator (SG) replacement, preventative maintenance procedures, *etc.* determine the degree to which capacity factors can be improved; shorter annual maintenance periods (decreased from 80 to 40 days), along with nearly zero unscheduled outages, make a 85-90% plant availability goal seem attainable. Plant life extension programs are evolving according to a long-term program wherein each piece of major equipment is subjected to a safety review every ten years. These programs together are expected to reduce capital costs by 20%, decrease O&M costs by 20%

(by the year 2001), as well as to reduce FC costs related to front-end and back-end (reprocessing, recycle, radwaste disposal) processes. While many of these BDC costs are being internalized for NE in Japan, other areas of potential NE credits are being examined. For example, oil supplies for Japan amount to 35 days, and the cost of storage of these supplies for NE time equivalents (*e.g.*, ~3 years) should either be attached to FE or attributed as credit to NE. Similarly, at ~100 \$/tonneC, the incremental COE amounts to 27 mill/kWeh (assuming ~100 yen/\$), and the cost of collected or avoided CO_2, should exceed the cost of NE waste disposal. At present, Kansai is working under Joint Implementation agreements (*e.g.*, providing CO_2-mitigating technologies to Thailand and to Indonesia), rather than dealing with carbon permits *per sé*.

F. Bouteille, Executive Vice-President, Nuclear Power International

"The French and German Project of the European Pressurized Water Reactor (EPR)"

The ongoing Franco-German collaborative design of a European Pressurized Reactor (EPR) was described as an evolutionary rather than a revolutionary design approach. Both economic and safety attributes of this system are being enhanced: improved (double) containment, reduced core-melt probability, reduced O&M costs, and reduced human-error risks. These attributes are being achieved through simplified safety designs, multiple redundancy, increased (accident-response) grace periods, improved man-machine interfaces, PRA systematic analyses, and systems studies of shutdown state; these latter points define the overall Strategy #1. Strategy #2 entails the overall mitigation of severe accidents. Core melting would be prevented through highly reliable cooling systems, and both hydrogen recombiners and the spreading of core-melt area would mitigate core-melting consequences. The reactor is divided into four sectors to prevent migration of potential hazard. Finally, economic competitiveness is a first priority of the EPR design through: a) increased efficiency (SG economizers, optimized secondary side operating at higher pressure); b) reduced FC charges (60 GWd/tonneHM burnup); c) reduced O&M (shortened service times, four-fold redundancy, preventative maintenance); and d) increased plant availability (shorter refueling outages, reduced reactor-trip frequencies). The thermal power has been increased 15% from 4,250 MWt to 4,900 MWt while reducing the residual heat-removal system from four to two trains outside the containment while not compromising core safety; extra boronation and emergency feedwater capacity are required. Generally, the EPR experience has combined two industrial and operational experiences using SOTA designs to give a lower generation cost (COE), while limiting the increase in capital cost.

Annotator Summary of Session I: J. Paffenbargner, Electricity Expert, IEA

A great divide seems to exist between existing NPPs (S. Brewer and H. Maeda) and (near-future) NPPs (F. Bouteille and M. Doucet), differing perspectives are identified, and no "secret passage" in going from the present into this near-term future could be identified. A central question for NE seems to be one of renewal *versus* reconstruction. Nuclear energy must prove competitive in the market place, rather than through governments or other (NE) conference presentations. Generally, NPPs are in a position to compete economically, but the future markets (of CCGTs, distributed generation, deregulation, *etc.*) present real challenges for NE capital and FC costs. A first goal is to provide low generation costs (COE) for systems that can be deployed in sizes that the market will absorb.

SESSION II

NUCLEAR COMPETITIVENESS - NUCLEAR POWER PLANT EVOLUTIONS

Moderators: B. Viellard Baron, Director, Framatome.
S. T. Brewer, President, S. Brewer Enterprises, Inc. Connecticut.
Annotator: H. Freslon, Vice-President, Nuclear Operations, Framatome.

Dissertators: P. Lecocq, Director, EDF

"EDF Actions to Improve the Competitiveness of Nuclear Energy in the Future"

The theme of economic competitiveness was repeated once more in this presentation, but from an EDF perspective. The maintenance of a competitive edge for NE in France (*e.g.*, O&M and FC costs) is being assured through efficient fuel and resource management, competitive operational expenses (<2%/yr of capital costs), enhanced plant availability (85%), and control of overall operating costs (~$4B in 1997). Actions that assure these goals are achieved include accountability of all players, benchmarking between designs, mastering decentralized operations, and mastery of volume and content of all maintenance operations. It was claimed that NEs economic margin with respect to CCGT is dwindling, with the key being strict control of investment expenditures, since safety is a big issue for NE. The overall (EDF) annual-charge breakdown was reported as 58% for investment, 20% for operations, and 22% for FC costs. The overarching EDF approach has been and continues to be standardization throughout the program. Construction costs are being reduced by: a) increasing unit sizes and the review of technical requirements as they evolve with the design; b) implementation costs are reduced by a complete program management system that minimizes the gap between design and construction, thereby reducing construction time and related interest payments; c) operating costs are controlled by extending operating campaigns (to 24 months), thereby approaching plant availabilities of 90%; and d) FC costs are reduced through increased fuel burnup (60 GWtd/tonneHM) and insertion of slightly MOXed units (*e.g.*, integration with the French reprocessing philosophy while maintaining FC operational flexibility). Elements that facilitate the optimal "industrial conditions" that are necessary for achieving these multi-front cost reductions include unambiguous technological definition (EDF is her own AE) that can be efficiently implemented through multi-packaged contracting, open competition, controlled rate of construction, and use of highly standardized designs. Based on this French model: a) assets exist or can be created to assure competitive NE; b) replacement or installation of evolving systems requires continuation of these (above-described) means; and c) implementation of new systems is through a series approach that is based on groups of 8-10 units at a time that are installed within 50-60 month/unit construction periods.

G. Davis, Director, ABB Combustion Engineering

"Competitiveness of Nuclear Plants in a Deregulated Market Place"

The focus of this presentation on NE competitiveness was on the U.S., where deregulation is driving dramatic differences in economic factors. Since the domestic market for new NPPs is non-existent in the U.S., activities have focused on assuring the competitiveness of existing NPPs in this deregulated electricity market while looking to developing Asian markets. Korea (both North and South) represents the present market (for

System 80⁺ designs), and, in competition with Framatome, China represents the market of the future. The System 80⁺ ALWR has been under development for the last 12 years, has now been certified by the NRC, and the need arises to go beyond the System 80⁺ in preparation for competition in the U.S. market. In this market, independent power producers (IPPs) are making market decisions that are strongly influenced by the desire to make short-term profits under conditions of minimum risk. External costs and community impacts are barriers to be avoided in this IPP-dominated deregulated market where PUCs are no longer available to provide bailout services in event of poor judgments made without perfect foresight. Consequently, aversion to capital-intensive technologies has climbed to new heights, and this aversion is driven by projections of non-nuclear technologies having low generation costs. A recent EPRI study was cited wherein the generation costs for NE, coal, and natural gas were given as 47-45, 28-30, and 25-35 mills/kWeh, respectively. Since FC and O&M costs have been brought well under control for NE, the disparity in capital costs of 1.4, 1.0-1.1, and 0.35-0.40 \$/We for NE, coal, and natural gas, respectively, must be addressed. Methods for reducing NEs capital costs by ~35% include: a) application of computer technology to the design, testing, certification, and installation of plant components (it was claimed that NE in the U.S. has a long way to go in this regard, particularly in the area of "smart" components to reduce redundancy costs); b) the development and application of PRA as a design tool for process simplification; and c) development and implementation of a risk-based, performance-based regulatory process in the design of new NPPs, and not just for plant operation (as is present practice). Creation of these new process tools will require broad R&D programs involving both government and industry in order to develop the tools *per sé*, to implement changes to industry and regulatory infrastructures, and to develop new licensing procedures. The enabling conditions needed for these changes to take place include: a) convincing government, industry, and the public that NPP designs can be simplified while remaining safe or being made safer; b) collaborations among government, laboratory, and university sectors; c) the development of a new long-term commitment of government and industry; and d) the simultaneous continuation of working issues for the present fleet of reactors. While the euphoria needed for these changes to occur is lacking, it was pointed out that cost savings through PRA in design and changes in regulatory processes hold little hope; engineering is important, but it does not represent a major part of the cost problem. Instead, the savings reside in the equipment, and NE must develop less-specific, less-specialized equipment while continuing to perform its functions safely. Changes or shifts to risk-based, performance-based regulations may be glacial, but these changes are useful in reducing costs associated with extended construction times and/or outages. Lastly, it was noted that some component redundancy is needed to assure acceptable levels of plant availability, and that safety and reliability are closely connected; the answer to reduced costs, therefore, may not reside strongly with the development and use of "smart" components.

B. Barre, Vice President, CEA

"Contribution of Research and Development to the Nuclear Power Competitiveness"

The role of R&D in establishing the competitiveness of NE is multifaceted and includes: a) understanding the physics and chemistry of processes within the NPP into order to transfer and expand that understanding to models and computer codes for the assessment of margins; b) removing or circumventing stumbling blocks related to accident conditions or excessive waste generation; c) qualifying and demonstrating safe operation; d) training or educating; and e) exploring technological and operational breakthroughs. All of the R&D functions contribute to economic competitiveness through both capital and O&M costs: a) by determining the degree to which unit total capital costs, UTC (\$/We), can be reduced through (constrained) increases in power ratings; b) by exploring the impact of component longevities (aging control, safety upgrades) and plant availabilities (instrumentation, length of operating cycle *versus* associated outage times) on COE(mill/kWeh); and c) by resolving the impact of NDT, corrosion, or lower-activation materials on overall O&M costs and, hence, COE. Research and development activities related to component aging assessment and control include studies of fatigue/creep, corrosion, and radiation damage (*e.g.*, swelling, embrittlement) processes. Similarly, R&D activities on the FC range from enrichment (AVLIS), fuel and cladding design and fabrication (high burnup, MOX parity, robustness), reprocessing, and waste management (volume reductions, packaging, storage, and disposal). Tools needed to conduct these R&D activities include critical facilities, system loops, "nuclearized" instrumentations, irradiation reactors, and hot cells. All of these activities generally impact the COE by at most 2-3%.

R. W. Gale, Chairman and CEO, Washington International Energy Group

"Why Nuclear Is Competitive and How It Can Stage a Comeback?"

This highly optimistic and aggressively pro-nuclear presentation claimed that NE is the cheapest way to make electricity. Good management of NPPs is the most important success factor. The nominal 7% (below 1990 CO_2 emission rates) Kyoto agreement needs NE. Furthermore, all types of NE were claimed to be competitive. While environment and economics are areas of optimal contributions for NE, nuclear power must first be competitive. Hence, NE is "green power", and a key question is whether the customer is willing to pay for this "green power". Agreement within industry is that NE must continue. The new management of NPPs are strongly "for-profit" operations in an environment that should and will force inefficient operators out of the market. Generally, competition breeds consolidation in order to reduce costs; EDF owns ~16% of the world NPPs, followed by Tokyo Electric (2% of the world NPPs); the NE industry has a long way to go in this regard. This industry needs to continue safe and economic operation of existing NPPs, with much less focus on new orders. Also, it was claimed that less focus should be placed on the back-end of the fuel cycle, while settling the waste issue. Then, while the industry moves through a generational change, alliances should be built with the environmentalists to push NEs "green power" agenda. As this process moves forward, the trend will be for relentless cost cutting, plant and people downsizing, productivity increases, and increased market share. Simultaneously with reductions in capital costs and overhang, NPPs should be consolidated (too many small operators at present), along with a consolidation of vendors, while embracing competition and building bridges with environmentalists.

SESSION III

SOCIAL ACCEPTANCE (PART I)

Moderators: J. Panossian, Director, SFEN.
Y. Gagarinski, Deputy Director, Kurchatov Institute.
Annotators: J. De La Ferté, External Relations, NEA-OECD.
D. Kyd, Head of Press Service, IAEA.
E. M. Hanson, (Los Alamos National Laboratory)

Dissertators: P. Haug, Executive Director, DAF and KTG.

"Working Towards Public Acceptance in Germany: Past Achievement and Present Challenges"

Public acceptance (PA) is strongly dependent on the regional political structure, which in Germany is highly de-centralized (8 nationwide producer-distributors, 700 total). Germany also has the highest electrical production per unit; electrical power production in Germany is a strong economic success story. Over 70% of Germans feel that NPPs should continue, if not expand, because of this generally good record. The group represented by the speaker (IK = Informationkreis Kernenergie) is comprised of utility members, suppliers, and consultants, who for 20 years have tried to show that NE is a safe, environmentally acceptable source of energy and, independent of political parties and companies, has worked on the Federal level to deliver this message. This core message is that NE is safe, carbon-free, needed, economic, resources-saving, and contributes to energy independence. This message is targeted primarily at political and youth groups in Germany. Despite this strong (70%) approval rating for NE, the recent Federal election (Gerhard Schroeder's Social Democrats received 40.9% of the votes, the CDU/CSU coalition 35.1%, and the Green Party 6.7%), the new ruling coalition (Social Democrats + Green Party, for a 47.6% plurality) has pledged on some yet-to-be-determined time schedule to close all German NPPs, support facilities, and related international collaborations. The lessons learned from this IK group are: a) find allies; b) identify important target groups; c) gather political assistance at an early stage and keep it; and d) convince utilities of the need for these kinds of activities. Even after 20 years of successful advocacy and an exemplary NE-industry track record in Germany, the present situation could not be predicted. When asked of the future of NE in Germany, the speaker envisaged a spectrum of time scales for NPP closures, varying from a year or so (Green Party) to decades (dictated by legal, economic, and contractual practicalities). Noted was the fact that Sweden is still wrestling with its NPP-closure mandate 20 years after its passage.

S. Bjurstrom, Director, SKB, Sweden

"Nuclear Waste and Public Opinion"

This speaker agrees with R. W. Gale that the back-end of the fuel cycle should be played down from the public-opinion viewpoint, but one should be proud of the NE industry performance, particularly in comparison with that of other generators and keepers of toxic wastes. Important determinants of public attitudes concerning nuclear waste include: a) concern about the environment; b) general fear of large technical systems; c) lack of trust in industry; and d) concern for NE, concern about nuclear weapons (NWs), and related global, national, and local influences.
J. P. Chaussade, Senior Advisor, EDF

"Nuclear Communication: How to Misunderstand the Wants and Needs of Public Opinion"

Since December 1996, EDF has pursued a policy of openness and transparency with regard to the French people and NE. This policy as brought back to France public acceptance (PA) that has been eroding since the Chernobyl reactor accident. Generally, public acceptance of NE is determined by a balance between good and bad attributes of NPPs. Advantages are self sustenance, ability to export electrical energy, low energy costs, reduced atmospheric emissions, and a credible EDF image. Disadvantages are embodied primarily in nuclear waste and health effects. Public opinion is generally favorable around NPP sites. The media is generally negative, but has not had much of an impact on public opinion; the public has greater trust in EDF officials than in journalists. EDF advertisements reach 50% of the people, 5,000 lectures are given each year, and 300,000 NPP visits occur per year (6 million visits to date). Advertising campaigns are most important, which after much testing led to the slogan "75% of electricity is nuclear". These advertisements also invite the public to learn more of NE. Currently, 84% of the French public feel that a solution to the nuclear-waste problem must be found; too much time has been spent selling electricity and not enough in dealing with the waste problem. Even though air pollution has fallen steadily since 1986, thanks to NE, 63% of the public think that NPP contributes to GHG emissions; clearly, more work is needed. Presently, the (French) media is attacking the economic benefits being claimed for NE.

A. Howard, Senior Vice President, NEI, US

"Public and Policy Maker Support for Nuclear Energy"

According to this speaker, the issues of waste, deregulation, and global warming are coming together. Policy makers are examining NE in a new light. A recent NEI survey of college students indicated an upswing in acceptance of NE by this sample group. Interestingly, most interviewees thought their neighbor (also from the sampled population?) did not support NE, in spite of the majority support on this issue. The NEI survey not only indicated a significant shift towards a more positive disposition, but also that those expressing favorable opinions thought (mistakenly) that the majority of the U.S. public held negative views; key findings of this survey (of 1000 adults, ±3% margin of error) for which agreement exceeds 60% are listed below:

- Favor use of NE as one way to provide electricity (61%, June 1998);
- NE should play an important future role (67%, March 1997);
- NE will be important in meeting future energy needs (70%, May 1997);
- The option to build more NPPs should be maintained (62%, June 1998);
- Manufactures of U.S. NPPs should play a leading world role (74%, November 1995);
- New advanced-design U.S. NPPs should be considered for electricity generation in your area (72%, June 1998);
- It is a good idea to renew licenses to current NPPs that meet Federal safety standards (80%, June 1998);
- NE is clean and reliable (63% on both counts, November 1995);
- Reduce unneeded stockpiles of weapons plutonium by processing it into fuel and using it for electricity (81%, February 1997).

Generally, an image problem exists among the informed U.S. public, with more favorable responses being given if that public is better informed on GHGs. The NE industry must share the "good news" if benefits are to be derived; NE is non-polluting, economic, possible to improve O&M without loss of safety margins, *etc.* In addition to having this message go out locally, it was suggested that the news media be used to shape public policy, to "de-mystify" technology, and to work on the above-noted perception gap (with regard to neighbor's NE views).

Annotator Summary of Session III: J. de la Ferté, External Relationships, NEA/OECD.
D. Kyd, Head of Press Service, IAEA.
E. M. Hanson, Los Alamos National Laboratory

The future of NE depends most on public acceptance issues and depends strongly on non-technical factors. The reasons for this situation are historic, deep-rooted, and everlasting. Furthermore, societal decision making has undergone radical changes over the years, with citizens gradually gaining new powers. Generally, the process of decision making in the public realm is described by a two-way process between proponents and other stakeholders. Given that the economic battle is (can be) won, one must focus on the PA issue by developing and improving NEs safety record and elaborating on NEs benefits through open and transparent policies; if the public knows more, acceptance increases.

KEYNOTE SPEAKER: C. Pierret, French Secretary of Industry

The achievements of the French NE program were summarized in terms of key metrics: 57 NPPs (PWRs); 78% of French electricity in 1997, increased from 22% in 1973; 60% decrease in acid rain and 20% decrease in CO_2 (60% of this decrease is attributed to NE). France is a major exporter of nuclear electricity in the EC. No new NPPs will be required until the year 2010. The choice of NE for France was determined by considerations of cost, security, and environment. Nuclear energy and FE are competitive under base-load conditions. Three main issues characterize the present energy policy: waste disposal; transparency of the NE society and public acceptance; and retention of the infrastructure. Presently, 20 French PWRs are licensed to recycle MOX. Waste disposal is being studied under the Law of December 31, 1991, in a series of Underground Laboratories, and a decision will be made by the year 2006. While the overall issue of transparency and PA is acceptable, room exists for improvement through the appointment of an independent body to oversee operations and decisions. The EPR represents the primary area where competition will be enhanced through improved safety and technical performance. Generally, it is too early to make final choices, and all options will remain open to give EDF power to maintain leadership in Europe and to maintain a diverse and well-balanced program.

SESSION IV

SOCIAL ACCEPTANCE (PART II)

Moderators: G. Clark, Secretary General, Uranium Institute.
B. Pellaud, Deputy General Director, IAEA.
Annotator: M. Kratzer, Former Deputy Assistant Secretary of State.

Dissertators: R. Wilson, Professor, Harvard University

"Safety and Regulatory Issues"

This presentation focused on the distinction between safety *goals* and safety *regulations*. A goal should be used to check whether a given regulation is reasonable, but should never be used as a regulation. The goal should be used to explain the safety issue and procedure to the public, and a spread should always exist between a goal and the corresponding regulation to show that the utility can in fact meet the letter of the regulation. Hence, a core-melt frequency of $<10^{-4}$ /yr is a goal. If the goal and regulation are synonymous, then the public is given the impression that the reactor is not as safe as it should be. It was proposed to check the safety goal against ALARA conditions, and then be willing to enforce the rules and regulations with some flexibility (*e.g.*, make the punishment for an infraction fit the crime). Other interesting tidbits injected into this talk included: a) the cost per man-rem for the NRC far exceeds that for medical irradiations ($10-100/man-rem *versus* $1000/man-rem), and the reasons for these differences should be understood; b) the U.S. will never meet the Kyoto accords since coal is too plentiful and gas is very inexpensive; c) while the FBR will never be built because of the high availability of uranium, one should be built for testing; d) the NRC should be sued if it exceeds its authority; and e) the NE future may never come because the system (in the U.S.) is over-regulated. Some signs exist of changes in the NRC, in that their approach is shifting to one that is following performance-based, risk-based procedures.

B. Pellaud, Deputy Director for Proliferation, IAEA

"With Good Safeguards the Civil Nuclear Industry Does not Contribute to Nuclear Proliferation"

The end of the cold war will not change the safeguard/proliferation issues; the question remains - does the existence of safeguards give cause for concerns about proliferation. The central question is whether safeguards applied to civilian NE can be simplified. Presently, the IAEA has 600 people to carry out 2,000 inspections/year; 50% of this staff are inspectors, and an annual (IAEA) budget allocated for this activity is $100M/yr for the inspection of all facilities under safeguards - R&D facilities, FC facilities, and NPPs, in descending order of priority. This $100M/yr budget does not include country contributions. Legally speaking, any facility under safeguards must contain nuclear material (NM). It is firmly believed that the NPT and the political environment have been responsible for containing the number of NW states, helped by the presence of IAEA inspectors. More aggressive inspection protocols were necessary to deal with the Iraqi situation in order to detect clandestine use of NM. In the future, the efficiency of safeguarding material will reduce the cost from $3,000 to $1,000/SQ through the use of advanced technologies and the better use of staff; the use of digital/modem technologies have contributed significantly to reducing staff travel time and related costs. The newer protocol calls for increased transparency (more information, extend beyond merely collecting data), more aggressive inspections and information gathering, the collaborative use of intelligence organization, and generally dealing with more information and more access.

Conventional safeguards have been improved through less-frequent visits (now once *per* month for a MOX facility), and consideration is being given to categorizing the kinds of plutonium in setting inspection frequency and procedures (presently, spent fuel is checked every three months, with reductions in inspection frequency being likely). The question of maintaining safeguard standards under conditions of an expanding NE enterprise remains, particularly since the old inspection paradigm of "check but don't challenge" is being replaced by a more aggressive one. It is noted that at an annual budget of $100M/yr and a world NE generation of ~2,300 TWeh/yr, the cost of present-day safeguards amounts to only 0.04 mill/kWeh, so the cost impact of the security that results from safeguards is not significant (~25 times less than spent-fuel disposal).

E. Arthur, Deputy Director for Civilian Application, Los Alamos National Laboratory (talk given by **E. M. Hanson**, Acting Deputy Director of Engineering Science and Applications Division, Los Alamos National Laboratory)

"Plutonium and the Nuclear Future"

The positive image of plutonium as an unlimited source of energy that is easily unlocked through the absorption of a single neutron by relatively abundant uranium-238 isotope has historically been contravened by its high radiotoxicity rating and by its long-term proliferation potential when stored either above ground in spent fuel, separated form within a processing unit or in to-be-irradiated MOX fuel, or below ground in a long-term repository. The negative images dominate the view from the end of the 20th century, as this material accumulates in most OECD countries at a rate of ~70 tonne/yr without any prospects for fast-spectrum burners (FSB = FBR, IFR, or ATW) or serious back-end NM management system on the drawing boards. After more than two years of analyzing a range of long-term global nuclear futures under the auspices of the Los Alamos Nuclear Vision Project, likely conclusions based on most-probable scenarios include:

- Globally, NE demand will grow with a significant upswing occurring possibly during the latter 1/2 to 2/3 of the 21st century;
- Use of NE shifts from the developed to the developing countries, mirroring the expected growths in population, productivity, and demand for primary energy;
- From a global perspective, the penetration of breeder reactors is projected to be slight until well after the middle of the 21st century, with increased NE demand driven by carbon-tax increases in FE possibly pushing this introduction date forward;
- Under these conditions, inventories of plutonium in spent fuel will grow, even for low-NE-demand scenarios.

Hence, the back-end of the FC will be important for any nuclear scenario; plans for dealing with both legacy and accumulated NM must be implemented. A long-term strategy has been developed for dealing with both NM legacy and accumulations according to the following four central goals:

- Eliminate present and future stocks of plutonium while recovering a portion of the contained energy value;
- Keep as much plutonium as possible in an unseparated form that is protected by strong radiation fields during fuel-cycle, fuel-recycle, and transport operations;
- Reduce the amount of plutonium and other actinides discharged from the fuel cycle to geologic disposition;

- Effect in the long-term significant reductions in global plutonium to a level dictated by future energy demands and the eventual need for plutonium breeding.

A four-component fuel-cycle architecture is proposed for implementing this strategy: a) transnational ownership of uranium enrichment facilities; b) continued safety and economic improvement of LWRs (ALWRs, ELWRs); c) secure interim storage of used commercial fuel in International Interim Monitored Retrievable Storage (IIMRS) facilities; and d) an ensemble of reprocessing/actinide-burning/power-producing facilities collectively operated as an Integrated Actinide Conversion System (IACS). These protected and self-contained IACSs would receive used fuel, possibly provide MOX to LWRs, convert plutonium and higher actinides to useful energy, possibly transmute long-lived fission products, return uranium to enrichment plants, and deliver optimized high-level waste to permanent disposal facilities. This four-point planning strategy accommodates a range of NE demand growth scenarios through the use of ALWRs while limiting the inventories of proliferation-prone plutonium and deriving energy therefrom. The majority of the anticipated NE demand would be generated in these MOX-burning ALWRs, whereas a fleet of FSBs (as identified above) would support 8-10 ALWRs with MOX while limiting the buildup of actinides through burning in FSBs. In addition to the FSB, the closed IACS sites would include IIMRS facilities, dry (pyrolytic/electrolytic) processing facilities, and facilities for encapsulating short-lived fission products for shipment to off-site waste depositories. The main goal of the IACS is to provide MOX to the ALWRs, and (ideally) an actinide-free waste stream to repositories that contain no long-lived fissions products. Key attributes of this approach are: a) enhanced proliferation resistance (plutonium is always in a high radiation environment, until fissioned); b) a range of NE production systems can be developed outside the plutonium-constraining/containing IACS, including systems that bridge to the thorium-uranium fuel cycle; and c) a commonality of technologies can be developed for NM management in a situation that does not close options; and d) a quantitative roadmap can be constructed that describes both the reductions in all (*e.g.*, legacy, weapons, and interim) plutonium inventories, and the technological requirements and time tables, for a range of specific global NE demand scenarios.

Annotator Summary of Session VI:, M. Kratzer Former Deputy Assistant Secretary of State, U.S.

Generally, the development of nuclear weapons preceded peaceful applications of nuclear science and technology; nuclear weapons can spread to additional nations, even if NE is abandoned. While risks of deriving nuclear materials from the civilian fuel cycle for use in weapon is not zero, safeguards are in place to prevent such occurrences. Typically, small dedicated facilities, and not NPPs or associated fuel-cycle facilities, have been the source of nuclear material for all NW proliferation to date, and remain the most probable route to any future proliferation. When asked to name any disadvantages perceived for NE, only 1% of U.S. college graduates identified NEs weapons or proliferation potential from a list of ten concerns, with accidents rating the highest (40%). Furthermore, 80% favored the burning of surplus weapons plutonium in NPPs over geologic disposal after immobilization in glass. These data suggest that the public is ahead of many policy makers in this area. Furthermore, the MOX option for disposal of surplus weapons plutonium is the only one that is capable of attracting Russian support in meeting this important international security objective. While nonproliferation considerations have not led to official U.S. opposition to nuclear power (an option of "last resort" with respect to greenhouse gas mitigation), their effect on U.S. fuel-cycle policy has been and remains profound. The issue of plutonium recovery and recycle, however, cannot be separated from that of spent fuel and the miscategorization of this material as "waste" having a low proliferation risk. Concerns of growing and (geographically) dispersed inventories of plutonium in used fuel can be addressed only by recycling this valuable energy resource; consumption and production of plutonium must be brought into balance, while reducing the number of countries where plutonium is stored (*e.g.*, creation of IIMRSs). Reprocessing should be allowed only in countries having unquestioned nonproliferation credentials, while simultaneously using proliferation-resistant technologies that allow repeated recycle and complete consumption of the actinides. These goals apply equally whether NE grows or is (unlikely) phased out.

Safeguards through effective, complete and independent verification are essential to maintaining confidence in nonproliferation to the extent necessary for nuclear power and international nuclear trade to take place. In contrast to present U.S. policy (*e.g.*, once-through fuel cycle), the long-term goal of international nuclear policy should bring plutonium production and consumption into balance and reduce inventories of plutonium in all forms, including spent fuel; this approach is most resistant to proliferation, given that a strong safeguards regime is maintained. A case can be made in this regard for relying on safeguards that are more akin to investigation than to (classical) verification, depending on a strong presumption of compliance, absent indications to the contrary. Recent developments indicate the need for further improvements in detection of undeclared activities, but undue concentration on implementation could weaken classical safeguards as applied to declared activities; integration of these protocol measures should be evolutionary, to avoid weakening the overall safeguards system. Lastly, breaches of (regional) physical security that risk placing nuclear materials in the hands of unauthorized parties have international consequences, but related international interests are inadequately reflected in the current arrangements.

SESSION V

DURABILITY - RESOURCES

Moderators: J. P. Rougeau, Senior Vice-President, COGEMA.
E. Arthur, Deputy Director of Civilian Applications, Los Alamos National Laboratory.

Dissertators: B. Krakowski, Senior Staff Member, Los Alamos National Laboratory.
L. Bennett, Senior Officer, IAEA, Retired.
E. Bertel, Principal Administrator, NEA.

"Nuclear Fission for Safe, Globally Sustainable, Proliferation-Resistant, and Cost-Effective Energy"

To varying degrees, under varying priorities, and depending strongly on country/region, the advancement of nuclear energy must deal with four cardinal issues: *waste*, *proliferation*, *cost*, and *safety*. The degree to which sustainability can be achieved depends crucially on the resolution of these issues. While solutions are usually offered within a technological/economic framework, the future of NE is being determined by these issues largely through a social/cultural paradigm and a dwindling public acceptance of this technology. From a general perspective of energy consumption in terms of minimal requirements, needs, and constraints, a reduction of the enormous differences in *per-capita* primary energy consumption that presently exists across the globe must be accompanied by a reassessment of how far above minimal levels does "the good life" reside; bringing the entire world population of a world ~50 years hence to *per-capita* primary energy consumption levels of current affluence is not feasible, even with large advances in technological efficiencies whereby primary and secondary energies are transformed into final-energy or end-use services. While the worldwide status of NE attained over the last three decades reflects a significant organizational and technological achievement, the present slow down of NPP demand and deployment in developed countries at a time of growing opportunities in developing countries presents concerns related to the ability to sustain the complex infrastructure required for the safe implementation. In view of the complexity of this technology, the status of NE represents an enormous accomplishment, but uncertainties driven by a shifting customer base, market transformation, and diminished public acceptance cloud prospects. Key elements of the public acceptance issue are described, and pathways are suggested to improve acceptance of this technology. The future of NE ultimately will be dictated by the level of public acceptance, and the evidence generally points to a public response to the technologies that define NE that is "value-ladened and cultural in context". This condition has far-reaching implications for any effort to win increased societal acceptance of NE.

Approaches to dealing with the all-determining public-acceptance issue revolves around adhering to a dedicated plan for breaking with the past to the extend needed to open new pathways to the resolution of social-cultural barriers that impede technological-economic advancements. A range of technical approaches to the future of nuclear energy offer new directions that are key to improved public acceptance. These approaches begin with securing the NPPs presently operating, followed by a transition to an intermediate phase based on ELWRs, and then entering a regime that combines NPPs with fast-spectrum burners (FSBs) of all actinides and long-lived fission products. Finally, the long-term and global impacts and trade-offs for nuclear energy as viewed through the lens of a simplified energy-economics-environmental (E^3) model emphasize how the future for nuclear energy may be tied to present concerns over the growing use of fossil fuels and possible climatic impacts related thereto.

W. Martin, Chairman, Washington Policy and Analysis, Inc.

"Short-Term Security of Resources"

No serious problem related to energy resources could be identified; most of the conditions that might lead to problems in this area are longer term (5-10 years). In the short term, however, potential issues could arise related to: a) the Caspian - infrastructure is a growing problem; b) the Middle East - stability is an issue, but conditions seem to be improving; c) Natural Gas - depending on demand, prices will rise, although probably not in the short term; d) Uranium - more than enough resource to meet short-term demand, as well as "business-as-usual" growth scenarios in the longer term; e) Coal - an enormous resource, unlikely to be impacted in the short term by GCC constraints; and f) Deregulation - presently on a fast track, but hopefully will be constrained to an evolutionary rather than revolutionary implementation.

F. Gére, Scientific Director, Defense Studies Foundation, Paris

"Strategic Aspects of Energy Resources"

This presentation gave a political science view of conflict, the use of energy, and the role of NE. The basic strategy, as reflected in the Iraq/Kuwait war is to keep that which is vital for life and to use military force to retain strategic energy resources. Three possible causes for conflict in this (energy) area were given: a) *need* for energy in search of increased prosperity; b) *frustration* in the face of no alternatives; and c) *reaction* by force resulting from the sum of the first two drivers (*e.g.*, need + frustration) in an effort to shift the balance of sharing of vital energy resources. The situation for conflict over energy is described by limited resources, competition for use, and political tensions as not good. These resources will be reduced as population and economic productivity grow, with energy demand being particularly strong in Asia. The role of technology, regional and global financial "meltdowns", (Asian) floods of last summer, effect of economic growth, and resource availability remain largely unknown, but a number of countries are ready to use force to reclaim their right to vital resources. The region of instability in this regard is seen as moving from the Horn of Africa north into the Azerbijan region and into Eurasia. New incentives for the use of force to enlarge access to energy resource are seen, including the growing environmental risks associated with increased use of FE, which creates a new kind of trans-boundary national threat. In addition to regional pollution leading to conflict, it was seen as unlikely that the U.N. will intervene in the global GHG issue, particularly since the U.S. is a major CO_2 emitter. The role of NE could be one of easing tensions related to FE misuse *vis á vis* excessive GHG emission, as well as competition for regional FE resources. Specific sources of tension related to NE, however, were identified, including:

- Cross-border seizure of NPPs - unlikely that a "dangerous" NPP could lead to this level of conflict before other measures short of war would be invoked;
- Coveting of uranium resources - also unlikely, in view of the price of this resource and the distribution of this resource; uranium (or thorium) is not in the same category as carbon in the application of FE;
- Nuclear waste management - while a problem for NE, this issue is not viewed as a source of conflict, particularly if that problem is internationalized in the future *vis á vis* the institution of regional nuclear regimes, IIMRSs, *etc.*;
- Proliferation risk - a possible cause of conflict, but international regimes are already in place to deal with related problems (NPT); regional conflicts related to rogue nation or sub-national behaviors, however, cannot be ruled out.

Generally, if reliance on FE continues to grow, traditional problems of conflict can correspondingly be expected to grow, and NE can reduce tensions and the tendency to use force to resolve FE resource and/or misuse issues. While this role may create a good social image for NE, this technology must deal first with bothersome problems related to an anti-social image (waste, proliferation, safety, economics).

P. Beck, Consultant, Royal Institute for International Affairs.

"The Global Energy Situation in the Next Century and Its Implications on Nuclear Power"

Nuclear energy suffers from the "Bonzo" syndrome - the market failure of a cheap, pleasant, and nutritious dog food with depressingly low sales because the dog did not like it. Truth matters little; perceptions matter, and perceptions have consequences. Fossil-fuel resources are enormous (particularly if methane clatherates are added to the balance sheets), and NE must be competitive with this FE abundance. Furthermore, public concerns are driven by safety issues and a history of NPP maloperations, proliferation (rightly or wrongly, something must be done about it), nuclear waste (a most visible choke point for NE), questions of whether this technology is applicable or usable in developing regions of the globe, and the availability of international and monitored standards. These uncertainties demand flexibility, and NE must compete with RE and an environmentally restructed FE option(s). Today's NE technology may not be suitable for an expanded future role, and more must be done to understand these shortcomings. First and foremost must be the regaining of public confidence in NE and the definition of a firmer role for this technology in the world's E^3 (energy-economics-environmental) future. Uncertainty in that future demands the flexibility of an open and viable NE option; if the public-acceptance and economics issues are adequately addressed, NE has a large role beyond 2050, although doubts are registered as to whether the present NE technologies can meet such an expanded role. No matter the outcome for NEs future, however, significant R&D efforts and related collaborations among countries will be needed to share costs and to convince the publics and governments that NE is a worthy endeavor. Finally and most importantly, a strong dialogue with NEs opposition is required to understand each others position and to reply to legitimate criticisms, at least at a level where agreement on life's uncertainties and the need for flexibility and options is concerned. What would an acceptable NE industry look like from the viewpoint of the opposition? In this regard, the need for a high-level, non-NE world council or study group was identified to examine these problems, including responses to an NE phase out and the requirement to maintain an infrastructure to deal with that limiting scenario.

SESSION VI

DURABILITY - EFFECTS ON THE ENVIRONMENT

Moderator: J. P. Laurent, Vice-President for Environment, COGEMA.
Annotator: S. Latek, National Atomic Energy Agency, Poland.

Dissertators: K. Leyden, Unit Head Directorate 17, European Commission

"Energy Futures in a Sustainable World"

Early energy expectations foresaw a FE resource shortage, new generation technologies that would reduce costs considerably, and strong overall growth in demand. Generally, none of these expectations have come to past. Presently in the EU carbon intensities (kgC/GJ or kgC/kWeh) have diminished because of the growth in NE as a competitive (but not "too cheap to meter") and environmentally benign source of energy. Without major policy changes in the next 10 years, however, NE will be phased out by 2030. The softness or hardness of this future "landing" for NE is a major question of transition that must be answered in the near term.

S. Sandklef, President and CEO, Vattenfall Braensle, Sweden

"Nuclear Power and Sustainable Development"

This presentation agreed with previous views that if the basis of NE resides only in the mitigation of CO_2 emissions, NE will lose. Nuclear energy must contribute to sustainable development, as defined in 1987 by the Brundland Commission - to satisfy our needs without compromising the needs of future generations. To that end, criteria based on fuel availability, land-use constraints, waste and related environmental constraints, availability and competitiveness of the required conversion technologies, and the impact on long-term GCC must be evaluated. The uranium fuel resource is concentrated and abundant; presently 3-years supply is in stock, 9 years are available from weapons stockpiles, 70 years supply are known to be available in the ground, estimated reserves in the ground amount to 170 years, 340 years of resource are available in phosphates, and 60,000 years supply reside in the oceans at a cost of ~200 $/kgU_3O_8$ (similar to the price of coal today). A 1-GWe NPP requires 1-4 km^2 of land, compared to 20-50 km^2 for solar(PV), 50-150 km^2 for wind, and 4,000-6,000 km^2 for biomass. Nuclear waste is long-lived, but compact in volume, and the costs of these wastes have been already internalized. Fuel costs have decreased in real terms (like oil), but NE must work on reducing capital costs. Generally, NE meets most of the criteria of a sustainable energy source. A regulatory environment that evaluates and controls all energy sources on the basis of fully internalized costs is needed.

J. Van Geel, Director, ITE, Karlsruhe

"Innovative Fuel Cycles"

The 2025-2100 time frame was identified as required to put innovative FCs into practice. The driving forces behind the implementation of these innovative FCs include most of the issues enumerated in previous presentations: public acceptance, waste control, non-proliferation, economic competitiveness, improved safety. These drivers have been tempered recently by the new realities associated with the growing stockpiles of civil plutonium and the possible release of military plutonium to the civilian fuel cycle. The new FCs are centered around the incineration of plutonium and include fast reactor burners (passively safe, lead-cooled, power production), HTGRs (high-temperature, high thermal-conversion efficiencies, power production for both the electrical grid and for process heat), and accelerator-driven (subcritical) systems (ADS, burners of plutonium and actinides, pursued both in EU and U.S.). Advances have occurred also in uranium mining, enrichment technologies, and in reactor core layout and composition. Alternative approaches to plutonium disposition include subsurface disposal of vitrified forms, MOX fueling with plutonium inventory growths reduced to near zero, burning of plutonium embedded in inert (non-breeding) matrices in both LWRs and FSBs, use of high-content plutonium fuels, and ultimately the use of non-proliferating thorium-plutonium fuels after Radkowsky or other variations. The nuclear waste issue is being addressed through interim storage, the separation of long-lived fission products (perhaps for transmutation in FSBs), and selective mineralization. Reprocessing technologies are following a number of promising routes, including the partitioning of minor actinides and long-lived fission products, and the use of dry processing techniques based on group separations in molten-salt/liquid-metal pyrochemical and electro-refining processes. The overall strategy of innovative FCs includes high plutonium consumption rates, low waste generation (low inventories, high recycle, low reagent radiation damage), ultra-high burnup fuels (200 GWtd/tonneHM), and the eventual replacement sequence LWR → ELWR → ELWR/ATW → (Th,U)LWR/ATW suggested by Los Alamos National Laboratory. A call was made for the formation of a high-level council of experts to prioritize and solve these FC problems.

S. Machi, Deputy Director General, IAEA, talk given by P. Danesi of IAEA

"Application of Radiation and Isotopes for Sustainable Development"

A world population of 5.9 B, a growth rate of 80 M/yr, 850 M living under conditions of chronic malnutrition, 6 M children dying each year for want of an adequate diet, growing shortages of water - these are the initial conditions for launching programs along some future trajectory that might lead to sustainable development. The agriculture needed to sustain, if not improve, this condition is also the world's biggest source of pollution (ground waters contaminated by food residue, secondary pests, and pesticides). The role of radiation and isotopes in dealing with these problems and in helping to steer a course to a sustainable global future was described. Examples of these worldwide efforts on the part of the IAEA include:

- Sterilized Insect Techniques (SIT) to reduce insects without pesticides (increased food production in some areas by 25%);
- Production of disease-resistant plants using fewer chemicals by use of mutants (hops, pasta, cotton, rice);
- Diagnostics for Suspended Particulate Matter (SPM) generated in the course of FE production;
- Diagnostics for monitoring heavy-metal emissions from coal-fired power plants and related mining operations;
- Nuclear radiations to clean flue gases emitted in the course of fossil-fuel burning;
- Desulfurization of flue gases using radiation, leading to the generation of ammonium sulfate for fertilizer use.

Generally, NE contributes a number of bi-products to the solution of mankind's problems, particularly in developing regions of the world.

CONCLUDING PANEL

Moderator: R. Linkohr, European MP, President of European Energy Foundation

The Moderator is a Social Democrat, but supports NE. Following Hegel, "ideas rule the world, not facts," it is believed that the world cannot be convinced with scientific facts. Microelectronics, chemical industry, genetic engineering, German (military) action outside of Germany, *etc.* all have had initial negative reactions followed by opinions that have changed. While it is important to never stop talking about the facts, other unforeseen events usually change opinion; in the long run, however, it is hope that man becomes science-driven, and not a policy-driven creature.

S. T. Brewer, President, S. Brewer Assoc.

The central issue is whether NE can compete in a deregulated market by the year 2005, and in such a market how new NPPs can be financed. Can an (Adam) Smithian free market "get it right"? For decades the (U.S.) Federal government has been a "nag and a nanny" for NE, and NE has been "orphaned" by the loss of the JCAE (Congressional Joint Committee on Atomic Energy). The Federal government can no longer salvage NE, but one must somehow bridge the gap to the future through securitization of old NPPs, as describe above.

W. Martin, Chairman, Washington Policy Analysis, Inc.

Energy is a security issue. Nuclear energy does well if FE is forced to internalize its costs. At present, FE fuel prices are low (deregulation is *supposed* to lower prices), overall growth in demand for all kinds of energy is low, and the consumer naturally wants the lowest price in the short term. This argument or trend, however, does not "close"; lower energy prices lead to more dependency on oil and gas, the large coal resource is difficult to resist, and both environmental and energy-security issues arise. Environment and security are the key points here. In this regard, it was proposed to combine forces on this environment/security/energy issue between the Council of Foreign Relations and Chatham House (Royal Institute of International Affairs) to form an International Energy Forum of the kind proposed above by Peter Beck. Inadvertently and informally, *quid pro quo* relationships have developed as the U.S. provides a military presence to keep ME oil flowing, and both the Japanese and French maintain and advance the NE option.

D. Vignon, Chairman and CEO, Framatome

Economics has and continues to drive NE, even in the presence of opposing environmental concerns. Fossil energy can sell fully standardized plants, and, as long as NE can not do the same, NE will not have a chance in such a market. Issues of safety, regulation, and certification are generally local and are quite different than in other industries (*e.g.*, CCGT generators, aircraft, *etc.*). The NE industries need to consolidate; too many suppliers exist for this shrinking market. Because NE generates a national security issue, however, consolidation of NE is not easily done; each country wants to maintain its own industrial capability. Nevertheless, safety rules and licenses need to be unified and collaborations maintained or extended beyond the conference level (*i.e.*, the EPR project).

W. Schmidt-Kuster, Foratom, Belgium

While GHG/GCC remains an important long-term issue, NE should not use the CO_2 argument as a *raison d'etre*; under these circumstances, NE will surely lose. The assistance provided by NE in meeting Kyoto goals, however, is still an important argument for NE. The safety issue for NE is solvable, but NE has not so far been able to solve the waste issue. A repository must be implemented along the logical lines followed by Sweden; projects rather than ideas are needed; we should not base our future on paper studies (*e.g.*, "paper-moderated, ink-cooled reactors"). Instead, we should make what we have work satisfactorily to answer NEs critics. Generally, the NE advocates are still talking too much among themselves and should re-focus communication with the public.

C. P. Zaleski, Deputy Director, CGEMP

Present NE economics look good; NPP life extension looks good; but competitiveness of new NPPs with regard to new FE generation does not look good. Only through the near-term intervention of governments in both NE and RE will the Kyoto plans be met (if ever implemented). A strong complementarity between NE and RE was noted: NE has a small (plant) "footprint" and extensive resources, and RE generally has a large "footprint" and rate-limited resources; both at present reside at opposite poles of the public acceptance issue, and NEs primary near-term challenge is to extend the consensus that resides at this conference to the rest of the population.

Closing Comments of Note from Both Panel and Non-Panel Participants:

B. Kursunoglu:

The separate issues of environment and proliferation are not parallel; one is positive and one is negative.

W. Schmidt-Kuster:

Proliferation is not a major issue in Europe, and the NE community should get on with other issues. Also, R&D on the effects of low-levels of radiation and on the application of transmutation technologies (ATW, ADS) are two areas of strong payoff and should be pursued vigorously.

P. Beck:

Proliferation is an issue in the U.S., and must be dealt with.

M. Kratzer:

Proliferation is not an issue in the U.S.

G. Clark:

A need exists to deal with the low-level radiation (effects) issue. Also, proliferation is an issue in the U.S. as a means to exert imperial control of a superpower.

L. Bennett:

Tritium production in a commercial NPP has been proposed, and the implications for the proliferation issue with regard to NE is questioned.

R. Wilson:

Also, accelerators are being proposed at Los Alamos for use in producing tritium for NWs and for transmuting waste products generated as part of the civilian NE program.

J. P. Rougeau:

With respect to the recent German vote and the vow to phase-out NE in Germany, and the relationship to the Franco-German joint venture on EPR, a contract is a contract, and the full force of that contract will be enforced with Germany. Also, working with the customers becomes very important as the energy markets globalize

R. Linkohr:

Transmutation (ATW, ADS) could form a bridge between advocates of a NE phase-in and advocates of an NE phase-out; such devices are needed in either case, and Europe should cooperate/collaborate in this area.

NOMENCLATURE

ABB	Asea Brown Boven
ADS	Accelerator Driven System
AE	Architect Engineer
ALARA	As Low As Reasonably Achievable
ALWR	Advanced Light Water Reactor
ATW	Accelerator Transmutation of (nuclear) Waste
AVLIS	Advanced Vapor Laser Isotope Separation
BDC	Beyond Direct Cost (externalities)
CCGT	Combined-Cycle Gas Turbine
CDU	Christian Democratic Union
CEA	Commissariat á l'Énergie Atomique
CEO	Chief Executive Officer
COE	Cost of Electricity (mill/kWeh)
CSU	Christian Socialist Union
D&D	Decontamination and Decommissioning
E^3	Energy-Economics-Environment
EDF	Electricité de France
ELWR	Evolutionary Light Water Reactor
EPA	Energy Policy Act (U.S., 1992)
EPRI	Electric Power Research Institute
EPR	European Pressurized Water Reactor
EU	European Union
FBR	Fast Breeder Reactor
FC	Fuel Cycle
FE	Fossil Energy
FPA	Federal Power Act (U.S., 1920)
FSB	Fast Spectrum Burner
GCC	Global Climate Change
GHG	Greenhouse Gas
HM	Heavy Metal
HTGR	High-Temperature Gas Reactor
IACS	Integral Actinide Conversion System
IAEA	International Atomic Energy Agency
IEA	International Energy Agency
IFR	Integral Fast Reactor
IIMRS	International Intermediate Monitored Retrievable Storage
IK	Informationkreis Kernenergie
IPP	Independent Power Producer
JCAE	(Congressional) Joint Committee on Atomic Energy
LWR	Light-Water Reactor
ME	Middle East
MOX	Mixed (uranium, plutonium) Oxide
MP	Member Parliament
NDT	Non-Destructive Testing
NE	Nuclear Energy
NEA	Nuclear Energy Agency
NEI	Nuclear Energy Institute
NM	Nuclear Material
NPP	Nuclear Power Plant
NPT	Non-Proliferation Treaty

NRC	Nuclear Regulatory Commission
NW	Nuclear Weapons
O&M	Operation and Maintenance
OECD	Organization for Economic Cooperation and Development
OT	Once Through (LWR fuel cycle)
PA	Public Acceptance
PE	Primary Energy
PRA	Probabilistic Risk Analysis
PUC	Public Utility Commission
PUCHA	Public Utility Act (U.S., 1935)
PURPA	Public Utility Act (U.S., 1978)
PV	Photovoltaic
PWR	Pressurized Water Reactor
R&D	Research and Development
RE	Renewable Energy
SG	Steam Generator
SIT	Sterilized Insect Technique
SOTA	State Of The Art
SPM	Suspended Particulate Matter
SQ	Significant Quantity (~8 kg Pu)
TPE	Total Primary Energy
UN	United Nations
UTC ($/We)	Unit Total capital Cost
WEC	World Energy Council

CONTRIBUTION OF PR C.P. ZALESKI TO THE CONCLUDING PANEL OF THE GLOBAL FOUNDATION 1998 MEETINGS

Pr. C.P. Zaleski

Centre de Géopolitique de l'Energie et des Matières Premières
Université Paris Dauphine
Place du Maréchal de Lattre de Tassigny
75775 Paris Cedex 16 FRANCE

An important seminar on nuclear energy in changing world was organized by the European Commission in Brussels last week.

I would like to report here some, relevant to our debate, highlights of presentations and lively discussions which took place there ; in the session of which I was Chairman.

1. COMPETITIVENESS OF NUCLEAR ENERGY

a. For existing plants the competitiveness is generally excellent – Operation and fuel cost low - therefore the life extension of existing plants should be a major objective.

b. The competitiveness of new nuclear plants seems difficult and may be achieved only in few selected countries.

The increased free competition in electricity production sector, makes the situation even more difficult, especially when the competitor is gas. The differences in time needed to build the plant and to recover the capital, make the investment in nuclear plants problematic.

2. KYOTO OBJECTIVES AND ITS LONGER TERM IMPLICATIONS

Almost all scenarios under consideration are far from satisfying the Kyoto objectives and implications.

Only vigorous action by Governments to constraint the demand and favour renewable and nuclear energy may accomplish the objectives (see scenario C of IASA-WEC).

Preparing the Ground for Renewal of Nuclear Power, edited by
Kursunoglu *et al.* Kluwer Academic / Plenum Publishers, New York 1999.

In all scenarios when Kyoto objectives are taken into account, the nuclear energy plays a significant role. For 2.030 considering economy realistically, the flexible mechanisms (Emission Trading, Joint Implementation and Clean Development Mechanism) can not be avoided.

A nuclear scenario will add significant benefit in terms of pollution permits price and CO_2 emission.

Another question is : will the Kyoto objective be implemented ? An US participant reported that according to his information the present US Senate would not ratified the Kyoto agreement. If this would be the case, that will lead to a major effect on the implementation in near future of the agreement.

One may however note that the progressive increase of political influence of environmentally oriented groups, which has allowed the signature of Kyoto agreement, may represent a stable and fundamental trend. If this will be the case, then a future US Senate may ratify Kyoto agreement and today possible refusal may represent only a delay.

3. A STRONG ANALOGY AND COMPLEMENTARITY OF NUCLEAR AND RENEWABLE ENERGIES WAS UNDERLINED BY MANY PARTICIPANTS

It was stated that nuclear has a small negative impact on environment like renewables and much less than fossil fuels.

In addition, the nuclear if associated to breeders has almost inexhaustible resources of fuel. This statement was not contested by anyone in our group. We may however notice that our group was not really representative of the society.

The challenge for nuclear energy is to extend this consensus to the society at whole as it was said in this conference many times.

4. RESOURCES EXHAUSTION

Strong opinions were expressed by many that exhaustion of fossil fuel, and especially of oil, shall have no major impact on prices by 2.030... in the market environment, even if it is clear that in few more decades one shall experience the progressive exhaustion of oil and later of gas.

Another group of participants expressed a strong view that saving finite resources of oil and gas for future generation is an important goal, comparable to prevention of greenhouse effect. Indeed, the exhaustion of precious commodity is irreversible and an important change of the environment for future generations. Therefore, before the exhaustion occurs, some anticipation in managing those resources is appropriated.

We have here the general debate between those who do not want to sacrifice or constrain the present for an uncertain future and those who do not want to leave irreversible and important damages of environment to future generation as price of present convenience.

Index

MIX
Papier aus verantwortungsvollen Quellen
Paper from responsible sources
FSC® C105338

If you have any concerns about our products,
you can contact us on
ProductSafety@springernature.com

In case Publisher is established outside the EU,
the EU authorized representative is:
Springer Nature Customer Service Center GmbH
Europaplatz 3, 69115 Heidelberg, Germany

Printed by Libri Plureos GmbH
in Hamburg, Germany